METHODS IN MOLECULAR BIOLOGY

Series Editor
John M. Walker
School of Life and Medical Sciences
University of Hertfordshire
Hatfield, Hertfordshire, UK

For further volumes:
http://www.springer.com/series/7651

For over 35 years, biological scientists have come to rely on the research protocols and methodologies in the critically acclaimed *Methods in Molecular Biology* series. The series was the first to introduce the step-by-step protocols approach that has become the standard in all biomedical protocol publishing. Each protocol is provided in readily-reproducible step-by-step fashion, opening with an introductory overview, a list of the materials and reagents needed to complete the experiment, and followed by a detailed procedure that is supported with a helpful notes section offering tips and tricks of the trade as well as troubleshooting advice. These hallmark features were introduced by series editor Dr. John Walker and constitute the key ingredient in each and every volume of the *Methods in Molecular Biology* series. Tested and trusted, comprehensive and reliable, all protocols from the series are indexed in PubMed.

Whole Genome Amplification

Methods and Protocols

Edited by

Yafeng Ma

Ingham Institute for Applied Medical Research, UNSW Medicine & Health, School of Clinical Medicine, Liverpool, NSW, Australia

Humana Press

Editor
Yafeng Ma
Ingham Institute for Applied Medical Research
UNSW Medicine & Health, School of Clinical Medicine
Liverpool, NSW, Australia

ISSN 1064-3745 ISSN 1940-6029 (electronic)
Methods in Molecular Biology
ISBN 978-1-0716-5059-2 ISBN 978-1-0716-5060-8 (eBook)
https://doi.org/10.1007/978-1-0716-5060-8

This Humana imprint is published by the registered company Springer Science+Business Media, LLC, part of Springer Nature.
The registered company address is: 1 New York Plaza, New York, NY 10004, U.S.A.

Preface

Whole genome amplification (WGA) has undergone remarkable evolution since its inception, transforming from a niche technique into a cornerstone of modern genomics. Early methods, such as degenerate oligonucleotide-primed PCR (DOP-PCR), enabled limited amplification of genomic DNA. Later breakthroughs, including multiple displacement amplification (MDA) and hybrid approaches like MALBAC, dramatically improved uniformity, reduced bias, and expanded compatibility with minute or degraded samples. Today, WGA is indispensable for unlocking genetic insights from challenging sources, from single cells and circulating tumor DNA to ancient fossils and forensic traces.

This book, part of a protocol series dedicated to WGA protocols, provides step-by-step methodologies refined through years of innovation. It bridges the gap between theory and practice, offering researchers reproducible techniques tailored to diverse needs. The protocols within address upstream challenges (e.g., sample preparation, quality control) and downstream applications (e.g., next-generation sequencing, SNP detection), ensuring end-to-end utility. WGA's impact spans fields like cancer genomics (detecting subclonal mutations), prenatal testing (noninvasive aneuploidy screening), microbiome research (amplifying low-biomass communities), and more, demonstrating its versatility as a foundational tool.

This volume focuses on *the broad application of WGA in cancers, preimplantation genetic testing, and forensic science*, with contributions from pioneers in other fields. Each chapter includes troubleshooting advice, cross-platform comparisons, and real-world examples to maximize success in the lab.

We extend our deepest gratitude to the contributors and the global scientific community for pushing WGA's boundaries. As the field advances, we hope this resource empowers your work and inspires future innovations.

Liverpool, NSW, Australia *Yafeng Ma*

Contents

Preface ... *v*
Contributors ... *ix*

1 Whole-Genome Amplification for Circulating Tumor Cells ... 1
Masato Kojima and Eiso Hiyama

2 Whole-Genome Amplification on Single Circulating Trophoblast Cell ... 11
Imen Chakchouk and Ignatia B. Van den Veyver

3 A Tailing Method of Quality Control on MALBAC-Based Whole-Genome Amplification Products ... 25
Changyue Chen, Jing Li, and ZengHui Xu

4 Micromanipulation and Multiple-Strand Displacement Amplification for Forensic DNA Profiling of Single Sperm Cells ... 37
Glenn M. G. Theunissen and Richard Jäger

5 Whole-Genome Amplification of Single Circulating Tumor Cells from Mice Xenografts ... 51
Nuria Estévez-Gómez, Cristóbal Fernández-Santiago, Roberto Piñeiro, and David Posada

6 An Integrated Preimplantation Genetic Testing Approach with MALBAC-Based Whole-Genome Sequencing ... 71
Cheng Wan, Fangfang Gao, and Sijia Lu

7 Handmade Biopsies and Whole Genome Amplification for Genotyping and Genomic Selection of Bovine Preimplantation Embryos ... 83
Clara Slade Oliveira, Luiz Sergio de Almeida Camargo, and Naiara Zoccal Saraiva

8 Accelerating Genomics: Innovations in Analyzing Limited and Rare Samples ... 93
Ioanna Andreou, Markus Storbeck, and Jonathan Schaffer

9 Sweet Sorghum Microspores' Fractionalization, Individual Isolation, and Whole-Genome Amplification ... 119
Chandana Kurapati, Mustapha Ọláwuni, Svitlana Lymanska, and Ahmad N. Aziz

10 Assessing Amplification Quality and Bias in MDA Methods Through Comparative Analysis of Short-Read Sequencing ... 131
E. D. Lozano-Escobar, V. Mateo-Cáceres, C. Mayoral-Campos, and M. Redrejo-Rodríguez

11 Sequencing RSV Whole Genome Using a Long Amplicon-Based Method with Oxford Nanopore Technologies ... 149
Xiaomin Dong, Steven Edwards, Yi-Mo Deng, Clyde Dapat, Arada Hirankitti, and Ian G. Barr

12 Laser Microdissection and Near Single-Cell Whole Genome Amplification 165
Roberto Cruz-Flores, Jorge Cáceres-Martínez, and Arun K. Dhar

13 Whole Genome Amplification of Microbial DNA from Host-Depleted Clinical Samples 175
Carl Halford, Lluis Moragues-Solanas, Simon A. Weller, and Matthew Gilmour

14 Whole Genomic Amplification for Detecting Single-Gene Variants Causing Inherited Disease 187
Tania Rojas-Pérez, Miguel Ángel Alcántara-Ortigoza, Ariadna González-del-Ángel, Felipe Camargo-Díaz, and Esther López-Bayghen

15 Single-Cell Low-Pass Whole Genome Sequencing of Circulating Tumor Cells 203
Zhuo Wang and Qihui Shi

16 Single Oocyst Selection and Whole-Genome Amplification of *Cryptosporidium parvum* 215
Yuancai Chen, Kaihui Zhang, Jianying Huang, Huikai Qin, Zi Yan, Yin Fu, Junqiang Li, Rongjun Wang, and Longxian Zhang

Index *225*

Contributors

MIGUEL ÁNGEL ALCÁNTARA-ORTIGOZA • *Laboratorio de Biología Molecular, Torre de Investigación. Instituto Nacional de Pediatría, México City, México*
IOANNA ANDREOU • *QIAGEN GmbH, Hilden, Germany*
AHMAD N. AZIZ • *College of Agriculture, Tennessee State University, Nashville, TN, USA*
IAN G. BARR • *WHO Collaborating Centre for Reference and Research on Influenza, Royal Melbourne Hospital, at the peter Doherty Institute for Infection and Immunity, Melbourne, VIC, Australia; Department of Microbiology and Immunology, University of Melbourne, at the Peter Doherty Institute for Infection and Immunity, Melbourne, VIC, Australia*
JORGE CÁCERES-MARTÍNEZ • *Centro de Investigación Científica y de Educación Superior de Ensenada (CICESE), Ensenada, Baja California, Mexico*
FELIPE CAMARGO-DÍAZ • *Laboratorio de Investigación y Diagnóstico Molecular, Instituto de Infertilidad y Genética México SC, INGENES, México City, México*
IMEN CHAKCHOUK • *Department of Obstetrics and Gynecology, Baylor College of Medicine, Houston, TX, USA; Duncan Neurological Research Institute, Texas Children's Hospital, Houston, TX, USA*
CHANGYUE CHEN • *Department of Medicial Research, Labway Company, Shanghai, China*
YUANCAI CHEN • *College of Veterinary Medicine, Henan Agricultural University, Zhengzhou, China*
ROBERTO CRUZ-FLORES • *Agriculture Pathology Laboratory, School of Animal and Comparative Biomedical Sciences, The University of Arizona, Tucson, AZ, USA; Centro de Investigación Científica y de Educación Superior de Ensenada (CICESE), Ensenada, Baja California, Mexico*
CLYDE DAPAT • *WHO Collaborating Centre for Reference and Research on Influenza, Royal Melbourne Hospital, at the peter Doherty Institute for Infection and Immunity, Melbourne, VIC, Australia*
LUIZ SERGIO DE ALMEIDA CAMARGO • *Embrapa Dairy Cattle, Juiz de Fora, MG, Brazil*
YI-MO DENG • *WHO Collaborating Centre for Reference and Research on Influenza, Royal Melbourne Hospital, at the peter Doherty Institute for Infection and Immunity, Melbourne, VIC, Australia; Department of Microbiology and Immunology, University of Melbourne, at the Peter Doherty Institute for Infection and Immunity, Melbourne, VIC, Australia; School of Biomedical Sciences, The University of Western Australia, Perth, WA, Australia*
ARUN K. DHAR • *Centro de Investigación Científica y de Educación Superior de Ensenada (CICESE), Ensenada, Baja California, Mexico*
XIAOMIN DONG • *WHO Collaborating Centre for Reference and Research on Influenza, Royal Melbourne Hospital, at the peter Doherty Institute for Infection and Immunity, Melbourne, VIC, Australia; Department of Microbiology and Immunology, University of Melbourne, at the Peter Doherty Institute for Infection and Immunity, Melbourne, VIC, Australia*
STEVEN EDWARDS • *WHO Collaborating Centre for Reference and Research on Influenza, Royal Melbourne Hospital, at the peter Doherty Institute for Infection and Immunity, Melbourne, VIC, Australia*
NURIA ESTÉVEZ-GÓMEZ • *CINBIO, Universidade de Vigo, Vigo, Spain; Galicia Sur Health Research Institute (IIS Galicia Sur), SERGAS-UVIGO, Vigo, Spain*

CRISTÓBAL FERNÁNDEZ-SANTIAGO • *Oncomet, Health Research Institute of Santiago de Compostela (IDIS), Santiago de Compostela, Spain*
YIN FU • *College of Veterinary Medicine, Henan Agricultural University, Zhengzhou, China*
FANGFANG GAO • *Department of Collaborative Innovation Center, Yikon Genomics Co., Ltd, Suzhou, China*
MATTHEW GILMOUR • *University of East Anglia, Norwich, Norfolk, UK; Quadram Institute Bioscience, Norwich Research Park, Norwich, Norfolk, UK*
ARIADNA GONZÁLEZ-DEL ÁNGEL • *Laboratorio de Biología Molecular, Torre de Investigación. Instituto Nacional de Pediatría, México City, México*
CARL HALFORD • *Defence Science and Technology Laboratory, Porton Down, Salisbury, Wiltshire, UK; University of East Anglia, Norwich, Norfolk, UK*
ARADA HIRANKITTI • *WHO Collaborating Centre for Reference and Research on Influenza, Royal Melbourne Hospital, at the peter Doherty Institute for Infection and Immunity, Melbourne, VIC, Australia*
EISO HIYAMA • *Department of Biomedical Science, Natural Science Center for Basic Research and Development (N-BARD), Hiroshima University, Hiroshima, Japan*
JIANYING HUANG • *College of Veterinary Medicine, Henan Agricultural University, Zhengzhou, China*
RICHARD JÄGER • *Department of Natural Sciences, Bonn-Rhein-Sieg University of Applied Sciences, Rheinbach, Germany; Institute for Functional Gene Analytics, Bonn-Rhein-Sieg University of Applied Sciences, Sankt Augustin, Germany*
MASATO KOJIMA • *Department of Pediatric Surgery, Hiroshima University Hospital, Hiroshima, Japan*
CHANDANA KURAPATI • *College of Agriculture, Tennessee State University, Nashville, TN, USA*
JING LI • *Shanghai HonsunBio Technology Co. Ltd., Shanghai, China*
JUNQIANG LI • *College of Veterinary Medicine, Henan Agricultural University, Zhengzhou, China*
ESTHER LÓPEZ-BAYGHEN • *Departamento de Toxicología, Centro de Investigación y de Estudios Avanzados del Instituto Politécnico Nacional (Cinvestav), México City, Mexico*
E. D. LOZANO-ESCOBAR • *Department of Biochemistry, Universidad Autónoma de Madrid (UAM) and Instituto de Investigaciones Biomédicas Sols-Morreale (CSIC-UAM), Madrid, Spain*
SIJIA LU • *Department of Collaborative Innovation Center, Yikon Genomics Co., Ltd, Suzhou, China*
SVITLANA LYMANSKA • *College of Agriculture, Tennessee State University, Nashville, TN, USA*
V. MATEO-CÁCERES • *Department of Biochemistry, Universidad Autónoma de Madrid (UAM) and Instituto de Investigaciones Biomédicas Sols-Morreale (CSIC-UAM), Madrid, Spain*
C. MAYORAL-CAMPOS • *Department of Biochemistry, Universidad Autónoma de Madrid (UAM) and Instituto de Investigaciones Biomédicas Sols-Morreale (CSIC-UAM), Madrid, Spain*
LLUIS MORAGUES-SOLANAS • *Microbiology, Biology Department, Universitat de les Illes Balears, Palma, Balearic Islands, Spain; Quadram Institute Bioscience, Norwich Research Park, Norwich, Norfolk, UK*
MUSTAPHA ỌLÁWUNI • *College of Agriculture, Tennessee State University, Nashville, TN, USA*
CLARA SLADE OLIVEIRA • *Embrapa Dairy Cattle, Juiz de Fora, MG, Brazil*

ROBERTO PIÑEIRO • *Oncomet, Health Research Institute of Santiago de Compostela (IDIS), Santiago de Compostela, Spain*
DAVID POSADA • *CINBIO, Universidade de Vigo, Vigo, Spain; Galicia Sur Health Research Institute (IIS Galicia Sur), SERGAS-UVIGO, Vigo, Spain*
HUIKAI QIN • *College of Veterinary Medicine, Henan Agricultural University, Zhengzhou, China*
M. REDREJO-RODRÍGUEZ • *Department of Biochemistry, Universidad Autónoma de Madrid (UAM) and Instituto de Investigaciones Biomédicas Sols-Morreale (CSIC-UAM), Madrid, Spain*
TANIA ROJAS-PÉREZ • *Laboratorio de Investigación y Diagnóstico Molecular, Instituto de Infertilidad y Genética México SC, INGENES, México City, México*
NAIARA ZOCCAL SARAIVA • *Embrapa Dairy Cattle, Juiz de Fora, MG, Brazil*
JONATHAN SCHAFFER • *QIAGEN Inc., Frederick, MD, USA*
QIHUI SHI • *Institutes of Biomedical Sciences, Fudan University, Shanghai, China*
MARKUS STORBECK • *QIAGEN GmbH, Hilden, Germany*
GLENN M. G. THEUNISSEN • *Brandenburg Police, Brandenburg State Criminal Police Office, Forensic Institute – Department of Forensic Biology, Eberswalde, Germany*
IGNATIA B. VAN DEN VEYYER • *Department of Obstetrics and Gynecology, Baylor College of Medicine, Houston, TX, USA; Duncan Neurological Research Institute, Texas Children's Hospital, Houston, TX, USA*
CHENG WAN • *Department of Collaborative Innovation Center, Yikon Genomics Co., Ltd, Suzhou, China*
RONGJUN WANG • *College of Veterinary Medicine, Henan Agricultural University, Zhengzhou, China*
ZHUO WANG • *Institutes of Biomedical Sciences, Fudan University, Shanghai, China*
SIMON A. WELLER • *Defence Science and Technology Laboratory, Porton Down, Salisbury, Wiltshire, UK*
ZENGHUI XU • *Shanghai Cell Therapy Group Company, Shanghai, China*
ZI YAN • *College of Veterinary Medicine, Henan Agricultural University, Zhengzhou, China*
KAIHUI ZHANG • *College of Veterinary Medicine, Henan Agricultural University, Zhengzhou, China*
LONGXIAN ZHANG • *College of Veterinary Medicine, Henan Agricultural University, Zhengzhou, China*

Chapter 1

Whole-Genome Amplification for Circulating Tumor Cells

Masato Kojima and Eiso Hiyama

Abstract

Circulating tumor cells (CTCs) are cancer cells that dissociate from primary and metastatic sites. The "liquid biopsy" analysis of CTCs is becoming attractive due to the easy access to blood. CTCs are thought to be the source of tumor progression and metastasis; genetic analysis of single CTCs allows the identification and characterization of tumor heterogeneity and malignancy. In clinical settings, the analysis of single CTCs can provide insights into tumor evolution and lead to precision cancer therapy. Genetic analysis of single cells is restricted by the low amounts of DNA (6–7 pg); therefore, whole-genome amplification (WGA) is required for downstream genetic analysis. In this chapter, we describe in detail the protocols from isolation of CTC from blood to WGA and quality check for downstream genetic analysis.

Key words Circulating tumor cell, Whole-genome amplification, Liquid biopsy, Single-cell analysis, Precision medicine

1 Introduction

Single-cell analysis of tumor cells has advanced rapidly over the last decade and has provided new insights into tumor heterogeneity and malignant biology. Circulating tumor cells (CTCs) are cancer cells that dissociate from primary and metastatic sites and enter the circulation. The analysis of single CTCs is becoming attractive owing to the easy access from blood as a "liquid biopsy" and because CTCs are thought to be the source of tumor progression and metastasis [1]. Moreover, the analysis of single CTCs will allow the identification and characterization of tumor heterogeneity because CTCs are derived from all tumor sites. In clinical settings, repeated sampling of CTCs can be performed with minimal invasiveness, and the analysis of single CTCs can provide insights into tumor evolution and facilitate precision cancer therapy. Genetic analysis of single cells is restricted by the low amounts of DNA (6–7 pg); therefore, whole-genome amplification (WGA) is

Yafeng Ma (ed.), *Whole Genome Amplification: Methods and Protocols*, Methods in Molecular Biology, vol. 3003, https://doi.org/10.1007/978-1-0716-5060-8_1,

required for downstream genetic analysis. In this chapter, we describe in detail the protocols from isolation of CTC from blood to WGA and quality check for downstream genetic analysis.

2 Materials

2.1 CTC Isolation

2.1.1 Equipment

1. Centrifuge.
2. Vortex.
3. Pasteur pipette and aspirator.
4. Hemocytometer.
5. Microscope.
6. Hand tally counter.
7. FACS Aria II (BD Biosciences).

2.1.2 Reagents and Consumables

1. Blood collection tube containing ethylenediaminetetraacetic acid (EDTA).
2. On-chip T buffer (On-chip Biotechnologies) (= cell absorption inhibitor).

   Replacement available: Anti-Link (Allvivo Vascular, Inc), Blockmaste PA1080 (JSR Life Sciences), Aggrewell Rinsing Solution (STEMCELL).
3. 15 mL conical tube.
4. Phosphate-Buffered Saline (PBS) with 2% (w/v) Fetal Bovine Serum (FBS).
5. 10 mL centrifuge tube.
6. Ficoll-Paque PLUS separation medium (Greiner Bio-One).
7. 1.5 mL LoBind DNA tubes (Eppendorf).
8. PBS.
9. Human BD Fc Block (BD Biosciences).
10. Fluorescence-labeled anti-GD2 antibody (BD Biosciences) and CD90 antibody (BioLegend) for isolating neuroblastoma CTCs in our laboratory.
11. Fluorescence-labeled anti-CD45 antibody.
12. Fluorescence-labeled anti-CD235a antibody.
13. PBS with 2% (w/v) FBS containing DAPI (0.05 ug/mL) (DAPI solution).
14. 0.2 mL PCR tube.

2.2 Whole-Genome Amplification

2.2.1 Equipment

1. Vortex.
2. Centrifuge.
3. Thermal cycler.

2.2.2 Reagents and Consumables

1. REPLI-g Advanced DNA Single Cell Kit (Qiagen).
2. Tris-EDTA buffer (pH 8.0): 10 mM Tris-HCl, 1 mM EDTA.
3. 8 M lithium chloride solution.
4. 0.5 M EDTA solution.
5. 100% ethanol.
6. 70% ethanol.
7. Qubit dsDNA BR Assay Kit (Thermo Fisher Scientific),

2.3 Quality Check of WGA Samples

2.3.1 Equipment

1. Thermal cycler.
2. CFX-96 Touch Real-Time PCR Detection System (Bio-Rad).

2.3.2 Reagents and Consumables

1. 0.2 mL PCR tube.
2. QIAGEN Multiplex PCR Kit (Qiagen).
3. Primers for multiplex real-time PCR of eight cancer-related genes: BRAF, EGFR, KIT, KRAS, NRAS, PIK3CA, PTEN, and TP53. Primer sequences are summarized in Table 1 [2].
4. 96-well plate.
5. Brilliant III Ultra-Fast SYBR® Green QPCR Master Mix (Agilent Technologies).

2.3.3 Software

1. CFX manager (Bio-Rad).

Table 1
Primer sequences of 8 cancer-related genes

	Forward	Reverse
BRAF	TACTGCTCTTTCTTCTCCAACAC	CCTGATTGTATTTGAGATCTAGTAGGG
EGFR	CAGCCTTCTCCGTAATTAGCAT	TGACACAGATAATTGTCCCACAG
KIT	GGCATTGAGGAGGGATAGTAAAT	CTGAACAATTTGCTTGAATGTTGG
KRAS	GTGTTACTTACCTGTCTTGTCTTTG	GCCTTCTAGAACAGTAGACACAA
NRAS	AATGGAATCCCGTAACTCTTGG	GATGATGTACCTATGGTGCTAGTG
PIK3CA	AGGGCAAATAATAGTGGTGATCT	CAGCAATTACTTGTTCTGGTACAC
PTEN	CTTTCTCTAGGTGAAGCTGTACT	GGTTCATTGTCACTAACATCTGG
TP53	AAGAGAAGCAAGAGGCAGTAAG	CTTAGGCTCCAGAAAGGACAAG

3 Methods

3.1 CTC Isolation

1. Collect blood (5–10 mL) from the patient at bedside and store it in blood collection tube containing ethylenediaminetetraacetic acid (EDTA).
2. Centrifuge the blood collection tube at 1700 × *g* for 10 min at room temperature.
3. Remove plasma (*see* **Note 1**).
4. Coat the inside of a 15 mL conical tube with 1 mL of On-chip T buffer (*see* **Note 2**).
5. Add the blood to the 15 mL conical tube (*see* **Note 3**).
6. Add Phosphate-Buffered Saline (PBS) including 2% Fetal Bovine Serum (FBS) to the 15 mL conical tube and mix by vortex. The volume of the PBS is the same volume of removed plasma.
7. Coat the inside of a 10 mL centrifuge tube with 1 mL of On-chip T buffer.
8. Add 3 mL of Ficoll-Paque PLUS (Greiner Bio-One) to the centrifuge tube.
9. Carefully layer the blood mixed with PBS onto the Ficoll-Paque PLUS (*see* **Note 4**).
10. Centrifuge the mixture at 400× *g* for 20 min at room temperature.
11. Coat the inside of a new 10 mL centrifuge tube with 1 mL of On-chip T buffer.
12. Remove the upper layer, leaving the mononuclear cell layer (*see* **Note 5**).
13. Transfer the mononuclear cell layer to the new centrifuge tube (*see* **Note 6**).
14. Suspend the mononuclear cell layer to 10 mL with On-chip T buffer.
15. Centrifuge the suspension at 600× *g* for 10 min at room temperature.
16. Aspirate the upper layer by Pasteur pipette (*see* **Note 7**).
17. Loosen the cell pellet by tapping and repeat the **steps 13–15**.
18. Suspend the cell pellet with 1 mL of On-chip T buffer, and transfer the suspension to 1.5 mL LoBind DNA tubes (Eppendorf).
19. Take 10 μL of the suspension into a new 1.5 mL LoBind DNA tube and add 90 μL of PBS and mix gently.
20. Take 10 μL of the diluted cell suspension and apply to a glass hemocytometer.

21. Using a microscope, focus on the grid lines of the hemocytometer with a 20X objective.
22. Count the live cells using a hand tally counter.
23. Centrifuge the cell suspension at **step 17** at 300× *g* for 5 min at room temperature and remove supernatant.
24. Suspend the cell pellet with PBS and Human BD Fc Block (BD Biosciences) (PBS: 175 μL/1 × 10^7cells, Fc Block: 25 μL/ 1 × 10^7cells).
25. Incubate for 10 min at room temperature.
26. Add fluorescence-labeled GD2, CD90 antibodies for the isolating neuroblastoma CTCs (30 μL/1 × 10^7cells, respectively), fluorescence labeled anti-CD45 antibody for eliminating leucocytes (30 μL/1 × 10^7cells), fluorescence-labeled anti-CD235a antibody for eliminating erythrocytes (50 μL/ 1 × 10^7 cells) to the cell suspension.
27. Use IMR-32, a neuroblastoma cell line, for positive control. Prepare cell suspension according to **steps 23–25**.
28. Put the cell suspension in a cool (4 °C) and dark place for 20 min.
29. Centrifuge the cell suspension at 300× *g* for 5 min at room temperature and remove supernatant.
30. Suspend the cell pellet with 200 μL of PBS with 2% (w/v) FBS and loosen by tapping, and repeat **steps 26–27**.
31. Suspend the cell pellet with DAPI solution (1.5 mL / 1 × 10^7cells).
32. Isolate targeted cells presumed as CTCs (GD2+, CD90+, CD45-, CD235a-, and DAPI-) by multiparametric fluorescence-activated cell sorting using FACS Aria II (BD Biosciences) and collect each cell into a 0.2 mL PCR tube containing 4 μL of REPLI-g Advanced sc storage buffer (Qiagen).

3.2 Whole-Genome Amplification

Perform whole-genome amplification with REPLI-g Advanced DNA Single Cell Kit according to the manufacturer's instructions.

1. Prepare Buffer D2, a mixture of 1 M dithiothreitol (DTT) and REPLI-g Advanced Buffer DLB (0.25 μL of DTT and 2.75 μL of REPLI-g Advanced Buffer DLB per sample).
2. Add 3 μL of Buffer D2 to the cell suspension of **step 32** in Subheading 3.1. Mix the tube by vortex and pulse-centrifuge (*see* **Note 8**).
3. Incubate for 10 min at room temperature.
4. Add 3 μL of Stop Solution. Mix the tube by vortex and pulse-centrifuge. Store on ice.

5. Prepare master mix, mixture of H_2O, REPLI-g Advanced sc Reaction Buffer, and REPLI-g sc DNA Polymerase (9 μL of H_2O, 29 μL of REPLI-g Advanced sc Reaction Buffer and 2 μL of REPLI-g sc DNA Polymerase per sample). Mix the tube by vortex and pulse-centrifuge (*see* **Notes 9** and **10**).
6. Add 40 μL of master mix to 10 μL of denatured DNA (from **step 4**).
7. Incubate for 2 h at 30 °C using a thermal cycler (*see* **Note 11**).
8. Inactivate REPLI-g sc DNA Polymerase at 65 °C for 3 min using a thermal cycler.
9. Add 10 μL of Tris-EDTA buffer.
10. Add 30 μL of 8 M Lithium chloride solution and 6 μL of 8 M 0.5 M EDTA solution and mix by tapping and pulse-centrifuge.
11. Incubate for 30 min at room temperature.
12. Add 150 μL of 100% ethanol and mix by tapping and centrifuge at 1500× *g* for 2 minutes at room temperature.
13. Remove the supernatant.
14. Add 100 μL of 70% ethanol and mix by tapping and centrifuge at 1500× *g* for 1 min at room temperature.
15. Repeat **steps 13–14**.
16. Remove the supernatant and pulse-centrifuge and remove residual ethanol.
17. Incubate for 10 min at room temperature.
18. Suspend the WGA sample with 50 μL of Tris-EDTA buffer. Store at 4 °C over night.
19. Measure the concentration of the WGA sample by Qubit dsDNA BR Assay Kit (Thermo Fisher Scientific) (*see* **Note 12**).

3.3 Quality Check of WGA Samples

1. Prepare reaction mix in 0.2 mL PCR tube, mixture of 2 × QIAGEN Multiplex PCR Master Mix (QIAGEN Multiplex PCR Kit), primer mix (2 μM each) and RNase-free water (7.5 μL of 2 × QIAGEN Multiplex PCR Master Mix, 1.5 μL of primer mix, and 5 μL of RNase-free water per sample). Mix the reaction mix gently but thoroughly, for example, by pipetting up and down a few times.
2. Add 1 μL of template DNA (80 ng/μL) to the reaction mix (14 μL/sample).
3. Carry out multiplex PCR using a thermal cycler. The reaction conditions are as follows: 95 °C for 15 min, denaturation at 94 °C for 30 s, annealing at 64 °C for 30 s, and extension at 72 °C for 30 s for a total of 10 cycles, with a final step of 72 °C for 10 min.

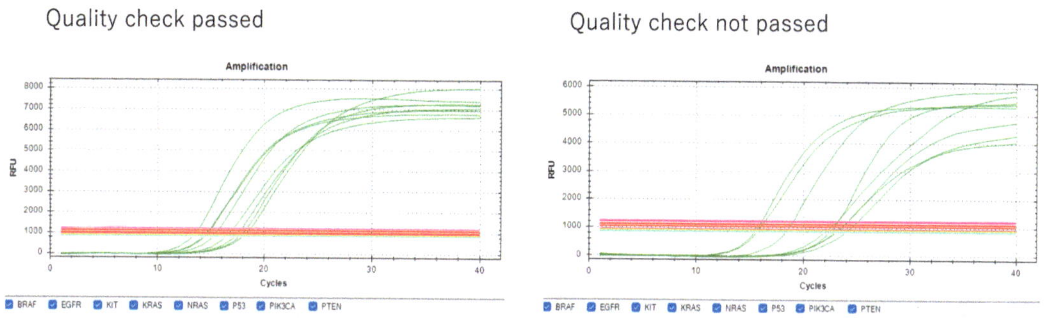

Fig. 1 Quality check of WGA samples. Quality check of WGA samples is performed by multiple qPCRs for eight cancer-related genes: BRAF, EGFR, KIT, KRAS, NRAS, PIK3CA, PTEN, and TP53. WGA samples with threshold cycles (Ct values) ≤ 20 at RFU 1000 in all eight cancer-related genes are classified as passing the quality check

4. Prepare reaction mix for real-time PCR, mixture of 2 × Brilliant III Ultra-Fast SYBR® Green QPCR Master Mix (Agilent Technologies), forward primer and reverse primer (10 μM each), and RNase-free water (5 μL of 2 × Brilliant III Ultra-Fast SYBR® Green QPCR Master Mix, 0.2 μL of forward primer, 0.2 μL of reverse primer, and 3.6 μL of RNase-free water per sample). Mix the reaction mix gently but thoroughly, for example, by pipetting up and down a few times.
5. Add the reaction mix (9 μL/sample) and 1 μL of multiplex PCR product to each well of the 96-well plate.
6. Carry out real-time PCR using CFX-96 Touch Real-Time PCR Detection System (Bio-Rad). The reaction conditions are as follows: 95 °C for 3 min, followed by 40 cycles of 95 °C for 5 sec, 60 °C for 10 s.
7. Real-time PCR data are reviewed and analyzed using the CFX manager (Bio-Rad). PCR products are assessed using threshold cycles (Ct values); samples with Ct values ≤20 at 1000 RFU in all eight cancer-related genes are classified as passing the quality check. (Fig. 1) (*see* **Note 13**).

4 Notes

1. When you remove the plasma, do not forget to record the volume because the same volume of PBS will be used later.
2. When you coat the inside of a 15 mL conical tube with 1 mL of On-chip T buffer, do not forget to coat the lid of the tube to minimize cell loss.
3. Coating the pipette chip with the On-chip T buffer before treating blood is preferable for minimizing cell loss.

4. When you layer the blood onto the Ficoll-Paque PLUS, do not mix them.
5. When you remove the upper layer, leaving the mononuclear cell layer, care should be taken not to disturb the mononuclear cell layer.
6. After transferring the mononuclear cell layer to a new centrifuge tube, wash the wall of the centrifuge tube with 1 mL of On-chip T buffer to minimize cell loss.
7. Do not forget to broil the tip of the Pasteur pipette to avoid contamination.
8. When you add 3 μL of Buffer D2 to the cell suspension of **step 32** in Subheading 3.1, make sure the cell material does not stick to the tube wall above the buffer line.
9. When you prepare the master mix for WGA, add the master mix components in the order described in the section of Methods. After the addition of H_2O and REPLI-g Advanced sc Reaction Buffer, briefly vortex and pulse-centrifuge before adding REPLI-g sc DNA Polymerase.
10. The master mix should be kept on ice and used immediately upon addition of REPLI-g sc DNA Polymerase.
11. When you incubate the master mix using a thermal cycler with a heated lid, the temperature of the lid should be set to 70 °C.
12. When you measure the concentration of the WGA sample, wait for the diffusion of amplified DNA in the suspension and measure on the next day of the WGA reaction. The concentration of the WGA sample typically ranged from 200 to 400 ng/μL.
13. The success rate is about 50%. The criteria of quality check are designed for comprehensive genetic analysis. More looser criteria would be suitable for targeted genetic analysis and others. The length of the PCR product of each primer is almost the same, and the Ct values of each gene depend on each WGA sample.

Sanger and next-generation sequencing (NGS) are used to identify genetic variants in single cells after WGA. NGS can target several genes, usually relevant oncogenes and tumor suppressor genes, that are included in gene panels. Array comparative genomic hybridization (aCGH) is used to detect CNVs. Digital PCR is frequently used to screen for known variants.

Acknowledgments

We would like to thank Editage (www.editage.com) for the English language editing.

Conflict of Interest Statement The authors have no conflicts of interest to declare.

References

1. Kojima M, Harada T, Fukazawa T, Kurihara S, Saeki I, Takahashi S, Hiyama E (2021) Single-cell DNA and RNA sequencing of circulating tumor cells. Sci Rep 11(1):22864. https://doi.org/10.1038/s41598-021-02165-7
2. Sho S, Court CM, Winograd P, Lee S, Hou S, Graeber TG et al (2017) Precision oncology using a limited number of cells: optimization of whole genome amplification products for sequencing applications. BMC Cancer 17(1): 457. https://doi.org/10.1186/s12885-017-3447-6

Chapter 2

Whole-Genome Amplification on Single Circulating Trophoblast Cell

Imen Chakchouk and Ignatia B. Van den Veyver

Abstract

Whole-genome amplification (WGA) is a technique designed to amplify the entire genome from trace amounts of original template DNA, such as that from a single cell. This amplification provides sufficient DNA for downstream comprehensive genetic analysis. In cell-based noninvasive prenatal testing (cb-NIPT), single circulating trophoblast cells are isolated from the maternal bloodstream. Among the required steps in single trophoblast isolation are light paraformaldehyde fixation, permeabilization, and immunostaining, which are not compatible with all WGA methods. Following single trophoblast isolation and WGA, the amplified DNA is subjected to detailed SNP genotyping and next-generation sequencing for genetic testing. This protocol is a step-by-step guideline for WGA on circulating single trophoblasts isolated from maternal blood during pregnancy.

Key words Whole-genome amplification, Single circulating trophoblast, Maternal blood

1 Introduction

Whole-genome amplification (WGA) is a crucial technique in molecular biology that enables the amplification of the entire genome from traces of DNA. This method is particularly valuable when dealing with a limited number of cells, ensuring that there is enough genetic material for downstream analyses such as sequencing, genotyping, and other genomic studies.

The process of WGA has revolutionized single-cell genomics, allowing researchers to investigate the genetic makeup of individual cells with high accuracy. This is crucial for various applications, including cancer research, where it helps in understanding tumor heterogeneity, and in prenatal diagnostics, where it enables noninvasive testing on circulating cells. Several different chemistries are available for single-cell WGA [1], for example, the degenerate oligonucleotide primed polymerase chain reaction (DOP-PCR)

Yafeng Ma (ed.), *Whole Genome Amplification: Methods and Protocols*, Methods in Molecular Biology, vol. 3003,
https://doi.org/10.1007/978-1-0716-5060-8_2,

[2], multiple displacement amplification (MDA) [3], multiple annealing and looping-based amplification cycles (MALBAC) [4], multiplexed end-tagging amplification of complementary strands (META-CS) [5], and Linear Amplification via Transposon Insertion (LIANTI) [6]. These developments have greatly enhanced the accuracy and reliability of single-cell genomic analyses.

In cell-based noninvasive prenatal testing (cb-NIPT), single trophoblast cells are isolated from maternal blood [7]. According to this protocol, the cells are fixed with paraformaldehyde, permeabilized, and immune-stained to ensure the preservation and identification of target cells. Once isolated, the single trophoblast cells are subjected to WGA to obtain sufficient DNA for detailed single-nucleotide polymorphism (SNP) genotyping and next-generation sequencing (NGS). These techniques allow for the detection of genetic abnormalities such as aneuploidies and copy number variations (CNVs) with high resolution.

Based on an extensive comparison of various whole-genome amplification kits, a few groups evaluated their efficiency and ability to accurately detect dosage differences in sex chromosomes, aneuploidy, and copy number variations (CNVs) in non-fixed single cells [8] and in fixed single cells [9, 10]. After careful consideration of factors such as amplification fidelity, sensitivity, reproducibility, and the capacity to detect even subtle genetic variations and comparison to other WGA approaches [11], Normand et al. found that the PicoPLEX WGA kit (Takara Bio) is the most suitable for single trophoblast cells isolated from maternal blood, which is the protocol we currently use.

After WGA, the amplified DNA is used to prepare libraries for shallow coverage whole-genome sequencing (WGS) and can also be used for SNP genotyping. The results from these processes were subsequently validated against clinical data to ensure accuracy and reliability. This approach successfully detected all aneuploidies and identified copy number variations (CNVs) of various sizes [9].

2 Materials

- Prepare the Tris-HCl buffer using ultrapure water to ensure optimal results.
- For all WGA reactions, use ultrapure PCR-grade, nuclease-free water to prevent any contamination in the amplification process.
- It is crucial to diligently avoid any DNA contamination of all reagents to maintain the integrity and accuracy of the experimental outcomes. Work in the pre-PCR area, and move to the post-PCR area only after the amplification step, using dedicated

pipettes for each area. Ensure the positive control gDNA tube is kept tightly closed and away from the kit tubes and single-cell tubes.

- Meticulous attention to detail in these steps will help ensure the reliability and validity of your results.

2.1 Positive Control DNA

1. Tris-HCl buffer: 1 M Tris-HCl:weigh 12.11 g Tris-HCl and transfer to 90 mL ultrapure water. Mix and adjust pH with HCl (*see* **Note 1**). Make up to 100 mL with ultrapure water. Dilute 1 M Tris-HCl to 5 mM using ultrapure water. Filter the diluted bench solution with a 0.2 μm sterile filter. Store at 4 °C.
2. Control Human Genomic DNA (gDNA): Extract, purify, and store the gDNA in 5 mM Tris-HCl (pH 8.0). Verify the integrity of the gDNA to ensure it is not degraded by running it on a 0.8% agarose gel; the band should be tight with no smear.

2.2 Single Cell Collection

The WGA input cells can be various cell types (*see* **Note 2**). In our protocol, we use human single trophoblast cells, which are collected and processed according to the method described in [7]:

1. Venous blood draw from pregnant women: BD Vacutainer purple top EDTA tubes (McKESSON).
2. Fixation buffer: 5% paraformaldehyde in phosphate-buffered saline (PBS).
3. Lysis buffer: 0.12% Triton X-100 in PBS.
4. Magnetic Cell Separation: MACS separation buffer (Miltenyi Biotec, Cat#130-091-221) and anti-biotin magnetic microbeads (Miltenyi Biotec, Cat#130-090-485).
5. Cocktail of capture antibodies: Anti-TROP2 and Anti-HLA-G and 2 anti-EpCAM (Novus Biologicals, Cat#NB500-452B and NB11055298B, BioLegend, Cat#324216, and JEG774 developed inhouse) (*see* **Note 3**).
6. Positive trophoblast selection column: magnetic-activated cell sorting MiniMACS columns (Miltenyi Biotec, Cat#130-042-201) (*see* **Note 4**).
7. Intracellular Staining Permeabilization Wash Buffer: BD Perm/Wash buffer (BD Biosciences).
8. Immunostaining antibodies: anti-cytokeratin antibodies (BioLegend, Cat#628608 and eBioScience, Cat#53–9003-82) for positive selection, an antibody to the white blood cell (WBC) marker CD45 (BioLegend, Cat#368510) for negative selection, and DAPI nuclear stain (*see* **Note 5**).
9. Resuspension cell buffer: calcium- and magnesium-free PBS (Corning).

10. Cells scanning: CyteSlides (RareCyte, Cat#24–1068-000) to spread the sample. CyteFinder (RareCyte) for slide scanning.
11. Single Trophoblast Isolation: CytePicker module (RareCyte) with 40-micron diameter needle.

2.3 Whole-Genome Amplification Procedure

1. After testing several methods [9] and considering that the circulating trophoblast cells are fixed, permeabilized, and stained, the PicoPLEX WGA kit (Takara Bio Inc., Cat#R30050) was selected for the WGA procedure. This is a one-tube, three-step reproducible whole-genome amplification of DNA from a single human cell, or other picogram quantities of DNA. This method was found to produce the most robust amplification and most reliable genotype and copy number variation (CNV) calls (*see* **Note 6**).

2.4 Purifying and Quantifying WGA Products

1. The DNA Clean and Concentrator-25 kit (Genesee Scientific, cat# 11–305) is used for rapid desalting and purification of the WGA product. The DNA size limit for this kit is between 50 bp and 23 kb, and recovery can be as much as 25 μg. The anticipated WGA product yield is approximately 5–10 μg.
2. Product is eluted in 1.5 mL tube Eppendorf DNA LoBind tubes.

2.5 Additional Materials and Equipment

1. Nuclease-free PCR tubes with tight seals (Thermo Fisher Scientific, Cat#AB0620): Significant evaporation (greater than 5 μL) can occur during incubation in a Thermocycler. This evaporation can compromise the robustness and reproducibility of the reaction. Therefore, using well-sealed reaction vessels is essential to maintain accuracy and consistency of experimental results.
2. Hot-lid PCR thermal cycler: For optimal results, it is recommended to use two hot-lid thermal cyclers. Utilize the first thermal cycler for the pre-amplification incubation in a designated pre-PCR area, and the second thermal cycler for the amplification step in a separate post-PCR area.

3 Methods

- Conduct all procedures up to the amplification incubation in a pre-PCR space, ideally under a clean hood or PCR station.
- Before the procedure, treat surfaces and pipettes with UV light for 30 min to ensure a sterile environment.
- Thaw the PicoPLEX tubes immediately before each step and follow each step's mixing instructions carefully to prevent the formation of bubbles.

3.1 Single Cell Collection

Single trophoblast cells are collected and processed using the protocol described in [7] following the steps listed below.

1. After obtaining informed consent, collect 40 mL of venous blood from pregnant women in BD Vacutainer purple top EDTA tubes.
2. Fix the blood sample for 10 min in 5% paraformaldehyde.
3. Lyse the red blood cells by incubating the sample for 8 min in lysis buffer containing 0.12% Triton X-100 in PBS.
4. Centrifuge at 700 g for 15 min and remove the supernatant.
5. Wash the recovered pellet in 150 mL PBS.
6. Centrifuge at 700 g for 15 min and remove the supernatant.
7. Resuspend the pellet in 1 mL PBS and 3 mL of MACS separation buffer.
8. Incubate the sample with a cocktail of 4 μg of each four capture antibodies for 2 h at 4 °C.
9. Centrifuge and remove supernatant. Adjust the volume to 1 mL with MACS separation buffer and add 100 μL anti-biotin magnetic microbeads and incubate for 2 h at 4 °C.
10. Apply cell suspension onto magnetic-activated cell sorting MiniMACS columns to perform positive trophoblast selection. Unlabeled cells will pass through.
11. Pipette 2 mL MACS separation buffer onto the MS Column. Immediately flush out the fraction with the magnetically labeled cells by firmly applying the plunger supplied with the column.
12. Centrifuge and remove supernatant.
13. Resuspend the recovered cells in 1 mL BD Perm/Wash buffer and immunostain for 1 h at room temperature with two anti-cytokeratin antibodies for positive selection, CD45 marker antibody for negative selection, and DAPI for nuclear stain.
14. Centrifuge and remove supernatant. Resuspend cell pellet in calcium- and magnesium-free PBS.
15. Spread the sample onto CyteSlides in PBS, with approximately half a million cells per slide. The slides are then scanned on CyteFinder. Candidate trophoblast cells are manually selected based on specific criteria: positive cytokeratin staining and its pattern, absence of CD45 staining, and distinct nuclear morphology.
16. Use the CytePicker module to pick individual trophoblasts with a 40-micron diameter needle and deposit them into nuclease-free PCR tubes in 2 μL of calcium- and magnesium-free PBS.
17. Store all cells at −80°C until further processing.

3.2 Preparation of Negative Controls

Whole-genome amplification (WGA) on single cells is highly sensitive to contamination. Ensuring reagent purity by including various negative controls is essential for accurate results (*see* **Note 7**).

1. Negative control N1: Add 5 μL of PicoPLEX Cell Extraction Buffer (green cap) to a clean PCR tube (green cap).
2. Negative control N2: Add 5 μL of the same PBS Buffer that is used for single cell isolation to a clean PCR tube.

3.3 Preparation of Positive Control

Prepare the positive gDNA control last to avoid DNA contamination of the single cell tubes and the negative control.

1. Dilute the stock of control unamplified genomic DNA to 1 ng/μL in the 5 mM Tris-HCl solution, vortex for 30 s.
2. Prepare this DNA to a final concentration of 15 pg/μL by adding 3 μL of the 1 ng/μL solution to 197 μL of 5 mM Tris-HCl, and vortex for 30 s.
3. Start by adding 4 μL of PicoPLEX Cell Extraction Buffer (green cap) to a PCR tube.
4. Add 1 μL of the 15 pg/μL gDNA solution to the PCR tube.

3.4 Cell Lysis and DNA Extraction

The WGA reaction is performed in the PCR tube into which the single circulating trophoblast cell was isolated by the CyteFinder. At each subsequent step, reagents are added directly to this tube.

1. Add 3 μL of Cell Extraction Buffer (green cap) to each cell sample to achieve a total volume of 5 μL. Avoid disrupting the cell; always add the reagent to the tube side (*see* **Note 8**).
2. Always prepare a master mix of PicoPLEX components, then add the appropriate volume to each tube. The number of samples is calculated by adding the number of isolated single cell samples, the three controls, and an extra sample to account for any potential pipetting errors (*see* **Note 9**).
3. Prepare the extraction cocktail by mixing 4.80 μL PicoPLEX Extraction Enzyme Dilution Buffer and 0.20 μL cell extraction enzyme (Table 1). Mix well by pipetting (*do not vortex*).

Table 1
Extraction cocktail component

Extraction cocktail component	Cap color	volume per sample	X # of samples	= volume per cocktail
Extraction enzyme dilution buffer	Purple	4.80 μL		μL
Cell extraction enzyme	Yellow	0.20 μL		μL
Total volume		**5.00 μL**		**μL**

Table 2
Extraction thermocycler cycle

Cycle	Temp	Duration
1 cycle	75 °C	10 min
1 cycle	95 °C	4 min
1 cycle	26 °C	Hold

Table 3
Pre-amplification cocktail component

Pre-amp cocktail component	Cap color	Volume per sample	x # of samples	= volume per cocktail
Pre-amp buffer	Red	4.80 μL		μL
Pre-amp enzyme	White	0.20 μL		μL
Total volume		**5.00 μL**		**μL**

4. Add 5 μL of the extraction cocktail to each single cell tube and control tube.
5. In the pre-PCR space, incubate in the hot-lid PCR thermal cycler as per Table 2.

3.5 Pre-amplification

1. During incubation, prepare the Pre-Amp Cocktail, by mixing 4.80 μL PicoPLEX Pre-Amp Buffer and 0.20 μL Pre-Amp Enzyme (Table 3). Vortex to mix.
2. Briefly spin down tubes, then add 5 μL of Pre-Amp Cocktail to each tube.
3. Incubate in the hot-lid PCR thermal cycler as indicated in Table 4.

3.6 Amplification

1. Prepare the amplification cocktail by mixing 25 μL PicoPLEX amplification buffer, 0.8 μL amplification enzyme and 34.2 μL nuclease-free water. Mix well by pipetting up and down (do not vortex) (Table 5).
2. Briefly spin down tubes, add 60 μL of amplification cocktail to each tube, and mix by pipetting up and down.
3. It is preferred to use another thermocycler in a post-PCR room.
4. Amplify in the thermocycler in the post-PCR area as indicated in Table 6.
5. After this amplification step, do NOT open the tubes in the pre-PCR room to avoid any contamination (*see* **Note 10**).

Table 4
Pre-amplification thermocycler cycle

Cycle	Temp	Duration
1 cycle	95 °C	2 min
12 cycles	95 °C 15 °C 25 °C 35 °C 65 °C 75 °C	15 s 50 s 40 s 30 s 40 s 40 s
1 cycle	4 °C	Hold

Table 5
Amplification cocktail component

Amplification cocktail component	Cap color	Volume per sample	x # of samples	= volume per cocktail
Amplification buffer	Orange	25.00 μL		μL
Amplification enzyme	Blue	0.80 μL		μL
Nuclease-free water	Clear	34.20 μL		μL
Total volume		**60.00 μL**		**μL**

Table 6
Amplification thermocycler cycle

Cycle	Temp	Duration
1 cycle	95 °C	2 min
14 cycles	95 °C 65 °C 75 °C	15 s 1 min 1 min
1 cycle	4 °C	Hold

3.7 WGA Reaction QC and Visualization

1. Prepare a 1% agarose gel in Tris-Acetate-EDTA (TAE) buffer, (MP#cat TAE50X01), add ethidium bromide (EtBr) (Sigma-Aldrich, Cat#E1510) to a concentration of 0.5 μg/ml.
2. In new tubes, mix 8 μL of 50% glycerol (Sigma-Aldrich, Cat#G7893) with 5 μL of WGA products.
3. Run the mix on the gel (Fig. 1) and store the remaining WGA products at −20 °C (*see* **Note 11**).

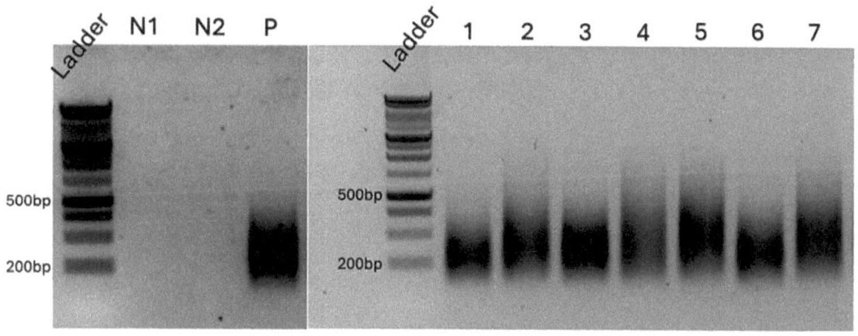

Fig. 1 Visualize the WGA product on a 1% agarose gel to assess the amplified products in lanes with negative controls, positive control, and WGA product from the isolated single trophoblasts (N1: negative control for cell extraction buffer; N2: negative control for PBS; P: positive control; 1–7: seven WGA products from seven different single trophoblast cells

3.8 Purifying and Quantifying WGA Products

The DNA Clean and Concentrator-25 kit is used for effective desalting and purification of WGA products, with an anticipated yield of 5–10 μg.

1. Add 5 volumes of DNA Binding Buffer to each tube with WGA product and mix by gently pipetting up and down (i.e., for a 50 μL WGA reaction, add 250 μL DNA binding buffer).
2. Let sit for 1 min at room temperature, then apply to a clean Zymo-Spin column with collection tube (be sure to label each column).
3. Spin the columns for 30 s at 15,000 rpm.
4. Add 200 μL Wash Buffer to the Zymo-Spin column (add the appropriate ethanol volume to the Wash Buffer bottle, as instructed on the label).
5. Spin 30 s at 15,000 rpm.
6. Add 200 μL Wash Buffer again.
7. Spin 60 s at 15,000 rpm.
8. Remove the spin columns and place each in a clean 1.5 mL tube DNA LoBind tubes.
9. Add 50 μL nuclease-free water directly to the spin column's membrane. Change pipette tips between columns!
10. Let the column sit for 5 min at room temperature.
11. Spin 30 s at 15,000 rpm to elute the DNA into the Eppendorf tube.
12. Label the Eppendorf tube clearly with sample ID, WGA method, date, and initials. Discard the column.
13. Measure WGA purified product concentration using Nano-Drop (Table 7) (*see* **Note 12**).
14. Store the samples at −20 °C.

Table 7
Nanodrop measurement of the WGA product of the positive control (15 pg gDNA) and seven single cells

#	Nucleic acid conc.	Unit	260/280	260/230
P	114.6	ng/μl	1.87	2.03
1	77.6	ng/μl	1.83	2.43
2	59.5	ng/μl	1.77	2.12
3	114	ng/μl	1.85	2.27
4	77.5	ng/μl	1.79	2.05
5	99.3	ng/μl	1.83	1.99
6	92.8	ng/μl	1.84	2.39
7	99.6	ng/μl	1.85	2.22

4 Notes

1. Concentrated HCl (12 N) can be used first to narrow the gap from the starting pH to the required pH. Then, use a series of HCl dilutions (e.g., 6 and 1 N) with lower ionic strengths to avoid a sudden drop in pH below the required pH.
2. Different input types of cells are suitable for WGA, including mammalian cells (1–10 cells), bacterial cells (1000–10,000 cells), or isolated DNA (6–50 pg of human DNA), intact or fragmented, single or double-stranded DNA. Input cells can be unfixed, fixed, or fixed/permeabilized.
3. Adjust the concentration of capture antibodies for each manufacturing lot. Based on the concentration of each of the four capture antibodies, prepare a cocktail sufficient for 12 samples, with each antibody contributing 4 μg per sample.
4. In this protocol, we used MiniMACS columns for positive selection. Alternative positive selection techniques can be used if they guarantee no cross-contamination between successive samples. For example [12], use a microfluidic sorting step for enrichment on a Namocell Namo 3 or Hana instrument, with each sample loaded on a separate flowcell.
5. This labeling antibody mix is developed for our enrichment and isolation protocol. If you are using another protocol, adapt the antibody choices [12, 13].
6. The PicoPLEX WGA kit (Takara Bio Inc., Cat#R30050) offers cell lysis and whole-genome preamplification, followed by

low-background amplification, resulting in over 2 μg of product. This protocol ensures accurate and reproducible measurement of CNVs. For other applications requiring longer reads or base-level resolution, other kits from Takara may be more suitable.

7. A pilot run can be conducted before the actual WGA, using both negative and positive controls. The intensity of the positive control band on the agarose gel electrophoresis ensures that all reagents are functioning properly, while clean negative control lanes confirm that the reagents are free of contamination.
8. The input volume of cells is very limited and not visible. The small volume containing the cells at the bottom of the PCR tube should not be disturbed by the pipette tip. If the tip touches this volume, there is a risk of losing the biological material, which could result in a failed reaction.
9. Number of samples = number of tubes with cells +1 positive control +2 negative controls +1 extra tube for pipetting error.
10. The amplification incubation should be performed in a different thermocycler located in a separate post-PCR room. WGA-amplified tubes should not be opened in the pre-PCR space to avoid contaminating future rounds of WGA.
11. The WGA is considered successful if the WGA product band is visible on the gel and distinct from the negative controls. The band intensity may vary depending on the quality of the input cells or DNA. Even if the band appears faint, the NanoDrop measurements will determine if the output product is sufficient for your subsequent applications.
12. If the NanoDrop concentration is as low as that of the negative control, the WGA reaction has failed. For our application, the concentration cutoff is set at 10 ng/μL. With this concentration, the maximum volume is needed as input volume for the next step of our protocol, which is NGS library preparation and shallow WGS. This yields a number of reads comparable to those obtained from WGA product concentrations of 30, 50, or 100 ng/μL.

Acknowledgments

This work was supported in part by grants R01HD055651 and P50HD103555 of the *Eunice Kennedy Shriver* National Institute of Child Health and Human Development of the National Institutes of Health and the Duncan NRI and research vision at Texas Children's Hospital. The contents of this publication do not necessarily reflect the views or policies of the NIH. The mention of trade names, commercial products, or organizations does not imply endorsement by the US Government.

Conflict of Interest Statement The authors have no conflicts of interest to declare that are relevant to the content of this chapter.

References

1. Huang L, Ma F, Chapman A, Lu S, Xie XS (2015) Single-cell whole-genome amplification and sequencing: methodology and applications. Annu Rev Genomics Hum Genet 16(1):79–102. https://doi.org/10.1146/annurev-genom-090413-025352
2. International HapMap C, Frazer KA, Ballinger DG, Cox DR, Hinds DA, Stuve LL, Gibbs RA, Belmont JW, Boudreau A, Hardenbol P, Leal SM, Pasternak S, Wheeler DA, Willis TD, Yu F, Yang H, Zeng C, Gao Y, Hu H, Hu W, Li C, Lin W, Liu S, Pan H, Tang X, Wang J, Wang W, Yu J, Zhang B, Zhang Q, Zhao H, Zhao H, Zhou J, Gabriel SB, Barry R, Blumenstiel B, Camargo A, Defelice M, Faggart M, Goyette M, Gupta S, Moore J, Nguyen H, Onofrio RC, Parkin M, Roy J, Stahl E, Winchester E, Ziaugra L, Altshuler D, Shen Y, Yao Z, Huang W, Chu X, He Y, Jin L, Liu Y, Shen Y, Sun W, Wang H, Wang Y, Wang Y, Xiong X, Xu L, Waye MM, Tsui SK, Xue H, Wong JT, Galver LM, Fan JB, Gunderson K, Murray SS, Oliphant AR, Chee MS, Montpetit A, Chagnon F, Ferretti V, Leboeuf M, Olivier JF, Phillips MS, Roumy S, Sallee C, Verner A, Hudson TJ, Kwok PY, Cai D, Koboldt DC, Miller RD, Pawlikowska L, Taillon-Miller P, Xiao M, Tsui LC, Mak W, Song YQ, Tam PK, Nakamura Y, Kawaguchi T, Kitamoto T, Morizono T, Nagashima A, Ohnishi Y, Sekine A, Tanaka T, Tsunoda T, Deloukas P, Bird CP, Delgado M, Dermitzakis ET, Gwilliam R, Hunt S, Morrison J, Powell D, Stranger BE, Whittaker P, Bentley DR, Daly MJ, de Bakker PI, Barrett J, Chretien YR, Maller J, McCarroll S, Patterson N, Pe'er I, Price A, Purcell S, Richter DJ, Sabeti P, Saxena R, Schaffner SF, Sham PC, Varilly P, Altshuler D, Stein LD, Krishnan L, Smith AV, Tello-Ruiz MK, Thorisson GA, Chakravarti A, Chen PE, Cutler DJ, Kashuk CS, Lin S, Abecasis GR, Guan W, Li Y, Munro HM, Qin ZS, Thomas DJ, McVean G, Auton A, Bottolo L, Cardin N, Eyheramendy S, Freeman C, Marchini J, Myers S, Spencer C, Stephens M, Donnelly P, Cardon LR, Clarke G, Evans DM, Morris AP, Weir BS, Tsunoda T, Mullikin JC, Sherry ST, Feolo M, Skol A, Zhang H, Zeng C, Zhao H, Matsuda I, Fukushima Y, Macer DR, Suda E, Rotimi CN, Adebamowo CA, Ajayi I, Aniagwu T, Marshall PA, Nkwodimmah C, Royal CD, Leppert MF, Dixon M, Peiffer A, Qiu R, Kent A, Kato K, Niikawa N, Adewole IF, Knoppers BM, Foster MW, Clayton EW, Watkin J, Gibbs RA, Belmont JW, Muzny D, Nazareth L, Sodergren E, Weinstock GM, Wheeler DA, Yakub I, Gabriel SB, Onofrio RC, Richter DJ, Ziaugra L, Birren BW, Daly MJ, Altshuler D, Wilson RK, Fulton LL, Rogers J, Burton J, Carter NP, Clee CM, Griffiths M, Jones MC, McLay K, Plumb RW, Ross MT, Sims SK, Willey DL, Chen Z, Han H, Kang L, Godbout M, Wallenburg JC, L'Archeveque P, Bellemare G, Saeki K, Wang H, An D, Fu H, Li Q, Wang Z, Wang R, Holden AL, Brooks LD, JE ME, Guyer MS, Wang VO, Peterson JL, Shi M, Spiegel J, Sung LM, Zacharia LF, Collins FS, Kennedy K, Jamieson R, Stewart J (2007) A second generation human haplotype map of over 3.1 million SNPs. Nature 449(7164): 851–861. https://doi.org/10.1038/nature06258
3. Dean FB, Nelson JR, Giesler TL, Lasken RS (2001) Rapid amplification of plasmid and phage DNA using Phi 29 DNA polymerase and multiply-primed rolling circle amplification. Genome Res 11(6):1095–1099. https://doi.org/10.1101/gr.180501
4. Zong C, Lu S, Chapman AR, Xie XS (2012) Genome-wide detection of single-nucleotide and copy-number variations of a single human cell. Science 338(6114):1622–1626. https://doi.org/10.1126/science.1229164
5. Xing D, Tan L, Chang CH, Li H, Xie XS (2021) Accurate SNV detection in single cells by transposon-based whole-genome amplification of complementary strands. Proc Natl Acad Sci USA 118(8). https://doi.org/10.1073/pnas.2013106118
6. Chen C, Xing D, Tan L, Li H, Zhou G, Huang L, Xie XS (2017) Single-cell whole-

genome analyses by Linear Amplification via Transposon Insertion (LIANTI). Science 356(6334):189–194. https://doi.org/10.1126/science.aak9787

7. Vossaert L, Wang Q, Salman R, McCombs AK, Patel V, Qu C, Mancini MA, Edwards DP, Malovannaya A, Liu P, Shaw CA, Levy B, Wapner RJ, Bi W, Breman AM, Van den Veyver IB, Beaudet AL (2019) Validation studies for single circulating trophoblast genetic testing as a form of noninvasive prenatal diagnosis. Am J Hum Genet 105(6):1262–1273. https://doi.org/10.1016/j.ajhg.2019.11.004
8. Zhang X, Liang B, Xu X, Zhou F, Kong L, Shen J, Xia Y, Xuan L, Mao Y, Xue Y, Liu C, Tan J (2017) The comparison of the performance of four whole genome amplification kits on ion proton platform in copy number variation detection. Biosci Rep 37(4). https://doi.org/10.1042/BSR20170252
9. Normand E, Qdaisat S, Bi W, Shaw C, Van den Veyver I, Beaudet A, Breman A (2016) Comparison of three whole genome amplification methods for detection of genomic aberrations in single cells. Prenat Diagn 36(9):823–830. https://doi.org/10.1002/pd.4866
10. Weymaere J, Vander Plaetsen AS, Tilleman L, Tytgat O, Rubben K, Geeraert S, Deforce D, Van Nieuwerburgh F (2020) Kinship analysis on single cells after whole genome amplification. Sci Rep 10(1):14647. https://doi.org/10.1038/s41598-020-71562-1
11. Zhang C, Zhang C, Chen S, Yin X, Pan X, Lin G, Tan Y, Tan K, Xu Z, Hu P, Li X, Chen F, Xu X, Li Y, Zhang X, Jiang H, Wang W (2013) A single cell level based method for copy number variation analysis by low coverage massively parallel sequencing. PLoS One 8(1): e54236. https://doi.org/10.1371/journal.pone.0054236
12. Bellair M, Amaral E, Ouren M, Roark C, Kim J, O'Connor A, Soriano A, Schindler ML, Wapner RJ, Stone JL, Tavella N, Merriam A, Perley L, Breman AM, Beaudet AL (2024) Noninvasive single-cell-based prenatal genetic testing: a proof of concept clinical study. Prenat Diagn 44(3):304–316. https://doi.org/10.1002/pd.6529
13. Afshar Y, Dong J, Zhao P, Li L, Wang S, Zhang RY, Zhang C, Yin O, Han CS, Einerson BD, Gonzalez TL, Zhang H, Zhou A, Yang Z, Chou SJ, Sun N, Cheng J, Zhu H, Wang J, Zhang TX, Lee YT, Wang JJ, Teng PC, Yang P, Qi D, Zhao M, Sim MS, Zhe R, Goldstein JD, Williams J 3rd, Wang X, Zhang Q, Platt LD, Zou C, Pisarska MD, Tseng HR, Zhu Y (2021) Circulating trophoblast cell clusters for early detection of placenta accreta spectrum disorders. Nat Commun 12(1):4408. https://doi.org/10.1038/s41467-021-24627-2

Chapter 3

A Tailing Method of Quality Control on MALBAC-Based Whole-Genome Amplification Products

Changyue Chen, Jing Li, and ZengHui Xu

Abstract

Single-cell whole-genome sequencing (scWGS) has illuminated new perspectives on genetic diversity across different cellular conditions. Despite this, the amplification of DNA from sparsely populated formalin-fixed tissues during the whole-genome amplification (WGA) process remains a formidable challenge. We introduce an enhanced quality control protocol for WGA products generated using the multiple annealing and looping-based amplification cycles (MALBAC) technique. This approach presents a cost-efficient, high-throughput quality control method, delivering a precise assessment of WGA efficacy with 100% accuracy on a test set of 128 samples.

Key words Whole-genome amplification, Quality control, MALBAC, Single-cell DNA sequencing, Low-input tailing method

1 Introduction

The burgeoning field of single-cell whole-genome sequencing (scWGS) has been significantly propelled by advancements in single-cell amplification techniques [1–7]. These methodologies are crucial for elucidating the genetic complexity of individual cells, offering unprecedented insights into the genetic heterogeneity present in both normal and pathological cellular states. The methods commonly employed include multiple displacement amplification (MDA), degenerate oligonucleotide primed PCR (DOP-PCR), and multiple annealing and looping-based amplification cycles (MALBAC) [8–10]. Among these, MALBAC stands out for its advantages in terms of uniformity, coverage breadth, and reproducibility, particularly when dealing with low cell number samples such as those derived from formalin-fixed tissues [8, 9, 11–15].

The MALBAC method is suitable for single cells obtained through flow cytometry sorting, buffer dilution,

Yafeng Ma (ed.), *Whole Genome Amplification: Methods and Protocols*, Methods in Molecular Biology, vol. 3003, https://doi.org/10.1007/978-1-0716-5060-8_3,

micromanipulation, and laser microdissection, as well as for micro-tissue cell samples that have been fixed with paraformaldehyde or subjected to laser microdissection (Fig. 1A). Amplification from minute amounts of DNA from microdissected samples usually has a problem with low genome coverage breadth [16], which restricts the exploration of tumor heterogeneity [17]. In prior studies, a multiplex quantitative polymerase chain reaction (qPCR) panel was utilized for quality control in whole genome amplification (WGA) processes [9, 18, 19]. In Navin's laboratory, a series of qPCRs were performed using 22 primer pairs independently [9]. WGA reactions that showed 20/22 positive qPCR reactions were selected for subsequent library construction and next-generation sequencing. However, the qPCR panel is not universally applicable in the quality control of MALBAC-based whole-genome amplification (WGA) due to the high input demands,vague cutoff thresholds, and complex operations [20]. This prompted us to investigate a new quality control (QC) method to evaluate WGA efficiency.

To address this, we have developed a novel, low-input tailing method for quality control of MALBAC-based whole-genome amplification (WGA) products based on the size distribution of MALBAC products. The tailing size represents the maximal size of a DNA fragment in capillary electrophoresis. When comparing tailing size with the peak at 10380 bp, tailings are divided into long tailing (≥10,380 bp peak) and short tailing (<10,380 bp peak). We demonstrated a strong correlation between the cutoff length of tailing method (10,380 bp) and 1× coverage breadth of 70% at 50× average depth (Fig. 1C).

2 Materials

Prepare all solutions using ultrapure water (prepared by purifying deionized water, to attain a sensitivity of 18 MΩ-cm at 25 °C) and analytical grade reagents. Prepare and store all reagents at room temperature (unless indicated otherwise). Diligently follow all waste disposal regulations when disposing of waste materials. We do not add sodium azide to regents.

2.1 Lysis Buffer

- 1 M Tris–HCl buffer, PH 8.8. Add about 800 mL of water to a glass beaker (*see* **Note 1**). Weigh 121.1 g Tris base and transfer to the beaker. Add water to a volume of 900 mL. Mix and carefully adjust the pH to 8.8 by adding HCl (*see* **Note 2**). Pour the solution into a 1 L volumetric cylinder and add water to bring the final volume to 1 L. Store at 4 °C.
- 100 mM EDTA (pH 8.0). Add approximately 80 mL of distilled water to a 100 mL glass beaker.Weigh 2.922 g of EDTA and add it to the beaker.Under vigorous stirring (using a magnetic stir bar), adjust the pH to 8.0 by slowly adding NaOH pellets or

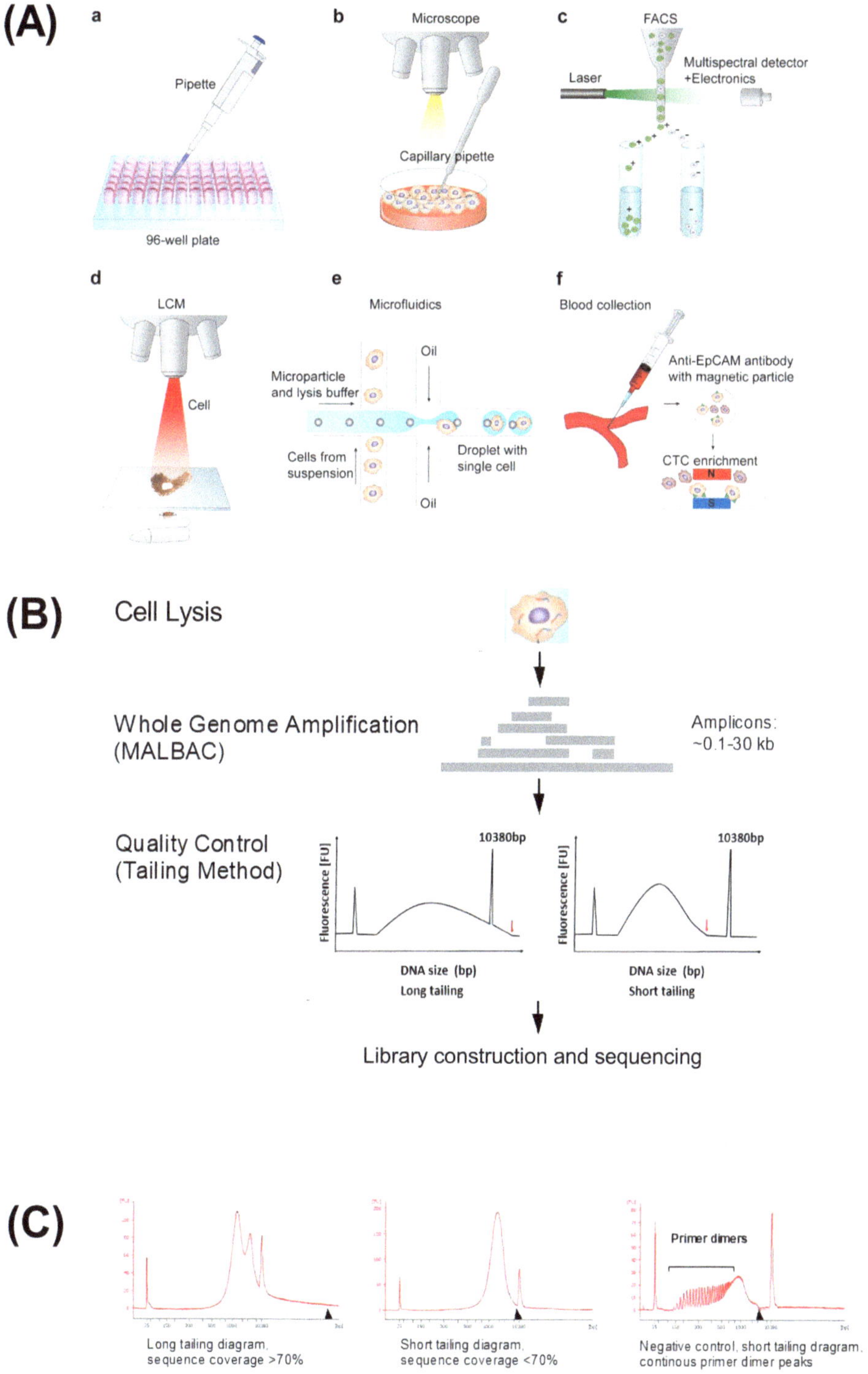

Fig. 1 (A) Single cells isolated from the following sources: (**a**) Buffer dilution. (**b**) Micromanipulation. (**c**) Flow cytometry sorting. (**d**) Laser microdissection. (**e**) Microfluidic technology. (**f**) The CellSearch system (paraformaldehyde fixation). (B) A schematic example of MALBAC amplification and quality control. Isolated cells were lysed and amplified by MALBAC methods and then proceeded with quality control using tailing method. Only

concentrated NaOH solution. EDTA will not dissolve completely until the pH approaches 8.0. Once fully dissolved, quantitatively transfer the solution to a 100 mL graduated cylinder. Rinse the beaker with a small amount of water and combine the rinses with the solution in the cylinder. Bring the final volume to 100 mL with distilled water and mix thoroughly.

- 100mM potassium chloride (KCl). Add approximately 80 mL of distilled water to a 100 mL glass beaker. Weigh 0.746 g of potassium chloride (KCl) and add it to the beaker. Stir or swirl until completely dissolved. Quantitatively transfer the solution to a 100 mL graduated cylinder. Rinse the beaker with a small amount of water and add the rinses to the cylinder. Bring the final volume to 100 mL with distilled water and mix thoroughly. Store at 4 °C.
- Lysis buffer: 0.03 M Tris–HCl, 2 μM EDTA, 20 μM KCl, 0.2% Triton, 40 μM DTT, 0.5 mg/mL Protease/Proteinase K. In a 50 mL sterile centrifuge tube, combine the following components in the order listed: 1.5 mL of 1 M Tris-HCl (pH 8.8), 1.0 μL of 100 mM EDTA (pH 8.0), 10.0 μL of 100 mM KCl, 1.0 mL of 10% Triton X-100, 2.0 μL of 1 M DTT. Add approximately 40 mL of sterile, distilled water to the tube and mix thoroughly by inverting. Using a sterile pipette, add 25 mg of Protease K powder directly to the solution. Gently swirl to dissolve and mix. Avoid vortexing to prevent foam formation and enzyme denaturation. Finally, add sterile, distilled water to bring the final volume to 50.0 mL. Mix by gentle inversion. Aliquot and store at -20 °C for long-term use. For each sample you intend to lyse, transfer 5 μL of the buffer solution to a well (*see* **Note 3**).

2.2 Pre-amplification Master Mix

- Pre-amplification NG/NT primer(10 μM:order NG primer 5′-GTGAGTGATGGTTGAGGTAGTGTGGAGNNNNNNGGG-3′and NT primer 5′-GTGAGTGATGGTTGAGGTAGTGTGGAGNNNNNNTTT-3′. After receiving the powder, dissolve it in 30 μL nuclease-free water, then quantify it with a NanoDrop, and continue to dilute it with nuclease-free water to a concentration of 10 μM.
- dNTP (10 mM concentration).

Fig. 1 (continued) the tailing (red arrow) of larger MALBAC amplicons (>10,380 bp) can pass quality control and proceed with library construction and sequencing. (C) Illustration of tailing method with electropherograms of 2100 Bioanalyzer. Samples with long tailing (black arrow) diagram correspond to >70% coverage breadth. Samples with short tailing (black arrow) diagram correspond to <70% coverage breadth. The negative control (without temple) with short tailing represents the feature of continuous primer dimer peaks and no large amplicons over 10,380 bp

2.3 Amplification Master Mix

- Amplification primer (10 μM): order NG primer 5′-GTGAGT GATGGTTGAGGTAGTGTGGAG -3′. After receiving the powder, dissolve it in 30 μL nuclease-free water, then quantify it with a NanoDrop, and continue to dilute it with nuclease-free water to a concentration of 10 μM.
- ThermoPol Buffer (10x concentration). Deep Vent (-exo) DNA Polymerase (5 U/μL concentration).

2.4 Purification of the MALBAC Products Using AMPure XP Beads

- AMPure XP beads (Beckman).
- Freshly prepared 80% ethanol: To prepare an 80% ethanol solution, mix 4.2 mL of 95% ethanol with 0.8 mL of distilled water in a clean container.

2.5 WGA Quality Control on Agilent 2100

- QuantiFluor® dsDNA Dye and Quantus™ Fluorometer (Thermo).
- NanoDrop (Thermo).
- Bioanalyzer High Sensitivity DNA kit (Agilent).

3 Methods

3.1 Cell Lysis

1. After isolated single cells or microtissue cell samples have been added to 5 μL cell lysis buffer (whether by pipette or cell sorting machine), firmly seal the plate with a cover to prevent evaporation of the sample during the thermal cycle process (*see* **Notes 4** and **5**).
2. Preheat the machine with a 50 °C Hold, before adding the plate to the machine.
3. Once the lid reaches 50 °C, add the plate and run the next rest of the program:

   50 °C for 20 min to 1 h (optimal temperature to enhance protease activity) (*see* **Note 6**).

   80 °C for 10 min (denatures protease).

   4 °C hold (this step preserves the DNA sample until it can be retrieved).
4. After the sample reaches 4 °C, it can be safely stored in a refrigerator at 4 °C for up to two months.

3.2 Pre-amplification Cycle for Single-Cell Whole-Genome Amplification (WGA) with MALBAC

1. Samples should already be lysed in 5 μL of lysis buffer (*see* **Note 3**).
2. Create pre-amplification master mix: For each sample, add 3 μL 10x ThermoPol buffer, 1 μL 10 mM dNTP, 0.75 μL 10 μM NG primer, 0.75 μL 10 μM NT primer. Make up to 25 μL with nuclease-free water.
3. Add 25.0 μL of the pre-amplification master mix solution to each of the samples (*see* **Note 7**).

4. Put the samples in the thermal cycler and run a denature cycle:

	1	2
Temp.	94 °C	94 °C
Time	05:00	Hold

5. Immediately after the denature cycle, quench the samples in ice.
6. Add 1 μL of Deep vent (-exo) DNA polymerase(5 U/μL) to each sample (*see* **Note 8**).
7. Put the samples in the thermal cycler and run the Pre-Amplification Cycle *8x times*:

	1	2	3	4	5	6	7
Temp.	10 °C	20 °C	30 °C	40 °C	50 °C	70 °C	95 °C
Time	00:20	00:40	00:40	00:40	00:40	3:00	00:20

8. Quench samples on ice immediately.

Move onto the amplification step when all eight runs of pre-amplification are completed.

3.3 Amplification (PCR) Step for Single-Cell Whole-Genome Amplification (WGA) with MALBAC

1. Create amplification master mix: for each sample, add 3 μL 10x ThermoPol buffer, 1 μL 10 mM dNTP, 1.5 μL 10 μM 27mer primer, 1 μL 5 U/μL Deep vent (-exo) DNA polymerase. Make up to 30 μL with nuclease-free water.
2. Add 24 μL of amplification master mix solution to each sample (use all pre-amplification product).
3. Put tubes/wells into the thermal cycler and run the Amplification Cycle (*see* **Note 9**):

		17–21x				
	1	2	3	4	5	6
Temp.	94 °C	94 °C	58 °C	72 °C	72 °C	4 °C
Time	00:30	00:20	00:30	03:00	05:00	Hold

Afterwards, DNA samples may be stored at −20 °C.

3.4 Purification of the MALBAC Products Using AMPure XP Beads

1. Ensure AMPure XP beads are removed from storage at least 30 min prior to starting the purification and fully equilibrated to room temperature, thoroughly resuspended by vortexing prior to use.

2. To each MALBAC products, add 77 μL (1.4X) of room temperature AMPure XP beads that have been thoroughly resuspended.
3. Once added, mix thoroughly and centrifuge briefly to collect all droplets. Do not allow beads to pellet (*see* **Note 10**).
4. Incubate the sample at room temperature for 5 min to allow the sample to bind to the beads.
5. Place the sample on a magnet to capture the beads. Incubate until the liquid is clear.
6. Carefully remove and discard the supernatant.
7. Keeping the sample on the magnet, add 200 μL freshly prepared 80% ethanol.
8. Incubate the sample at room temperature for ≥30 s.
9. Carefully remove and discard ethanol, without disturbing the bead pellet.
10. Keeping the sample on the magnet, add 200 μL freshly prepared 80% ethanol.
11. Incubate the sample at room temperature for ≥30 s.
12. Carefully remove and discard ethanol, without disturbing the bead pellet.
13. Allow the beads to dry at room temperature for 3–5 min, sufficiently for all remaining ethanol to evaporate (*see* **Note 11**).
14. Remove the sample from the magnet.
15. Thoroughly resuspend the beads in 25 μL of 10 mM Tris–HCl (pH 8.0–8.5) or nuclease-free water.
16. Incubate the sample at room temperature for 2 min to elute the sample off the beads.
17. Place the sample on a magnet to capture the beads. Incubate until the liquid is clear.
18. Transfer the appropriate volume of the clear supernatant to a fresh tube and proceed with DNA quantification.
19. Purified, amplified sample can be stored at +2 °C to +8 °C for 1–2 weeks or at −15 °C to −25 °C for up to 3 months.

3.5 WGA Quality Control on Agilent 2100

1. Quantify the purified MALBAC products with QuantiFluor® dsDNA Dye and Quantus™ Fluorometer or NanoDrop.
2. Dilute each MALBAC product to 1 ng/μL and analyze using a Bioanalyzer High Sensitivity DNA kit (*see* **Notes 12** and **15**).
3. Evaluate the size distribution and tailing size of MALBAC products (*see* **Notes 13**, **14**, and **16**). Only the tailing of larger MALBAC amplicons (>10,380 bp) can pass quality control and proceed with library construction and sequencing (Fig. 2).

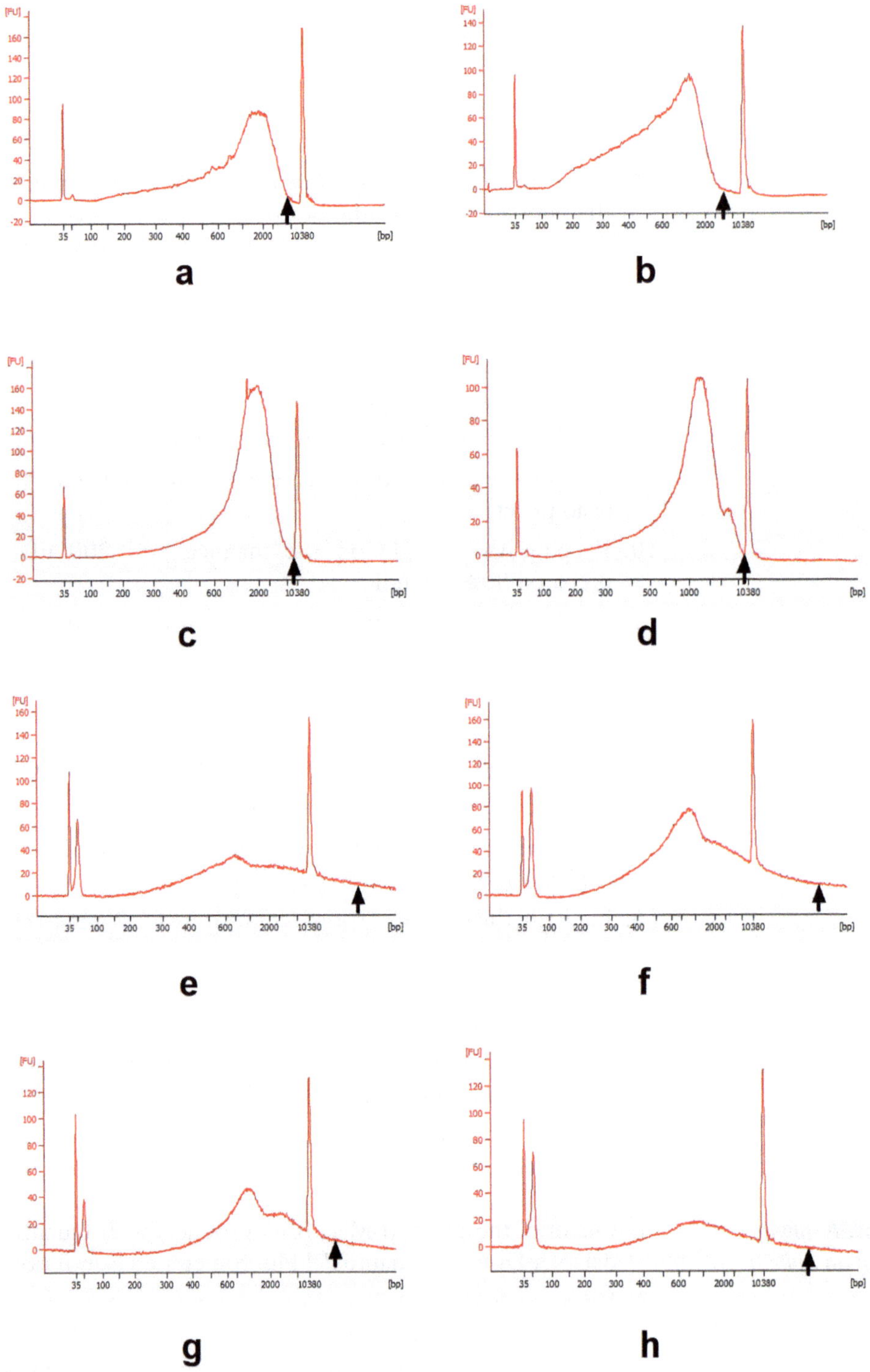

Fig. 2 Electropherograms of multiple annealing and looping-based amplification cycles (MALBAC) products. Short tailing group (a) 5952 bp (black arrow), (b) 5183 bp (black arrow), (c) 8701 bp (black arrow), (d) 8766 bp (black arrow), corresponding 1× coverage breadth of (a) 19.41%, (b) 23.50%, (c) 46.55%, (d) 66.32%. Long tailing group (e) 20,911 bp (black arrow), (f) 25,103 bp (black arrow), (g) 14,943 bp (black arrow), (h) 18,217 bp (black arrow), corresponding 1× coverage breadth of (e) 75.45%, (b) 79.37%, (c) 81.11%, (d) 87.52%

4 Notes

1. The use of warm water (~37°C) in a glass beaker significantly accelerates the dissolution of Tris base. A magnetic stir bar can be introduced immediately to facilitate this process. The solution must be cooled to room temperature prior to pH adjustment.
2. Concentrated HCl (12 N) can be used at first to narrow the gap from the starting pH to the required pH. From then on, it would be better to use a series of HCl (e.g., 6 and 1 N) with lower ionic strengths to avoid a sudden drop in pH below the required pH.
3. Create a master mix in an Eppendorf tube rather than make buffer for each sample separately. Prepare a single large amount of solution that will be enough for all the samples. Microscopic cells can easily be blown out of the tube, so be careful with the operation; add reagents along the wall of the tube without blowing.
4. To avoid contamination with exogenous DNA and the presence of amplification inhibitors during the cell preparation process, it is recommended to wash the cells before the experiment using a 1× PBS solution that is free of Mg^{2+} and Ca^{2+}. Please ensure that the volume of PBS solution used in subsequent experiments is carefully controlled and does not exceed 1 μL.
5. This method is suitable for whole-genome amplification of single cells obtained from the following sources: flow cytometry sorting, buffer dilution, micromanipulation, and laser microdissection, among others. Additionally, this kit is also applicable for micro-volume tissue cell samples that have undergone paraformaldehyde fixation or laser microdissection.
6. Proteinase K is a widely utilized lysing enzyme capable of breaking down cell membranes and walls, facilitating the release of cellular contents. The optimal conditions for Proteinase K treatment can vary significantly based on the sample type and the specific objectives of the experiment. For instance, fresh mammalian cells typically require only a 20-min incubation for effective lysis. In contrast, the denser cell walls of bacteria demand a more extended treatment, often up to 1 h. Furthermore, cells that have been fixed with formaldehyde or tissues derived from microdissection often require an extended duration, with a general recommendation of 1 h. It is also important to note that the structural complexity and cellular density of different tissues may influence the necessary Proteinase K concentration and treatment duration, necessitating tailored adjustments for optimal results.

7. To avoid sample contamination, thoroughly clean the workbench with specialized nucleic acid contamination prevention reagents; change gloves before preparing the mix and adding samples; To prevent contamination from the operator's shed cells; clean the trash bin after each experiment. Ethanol can inhibit enzyme activity to a certain extent, so if you use ethanol for wiping, please wait until the ethanol has completely evaporated before proceeding with the reaction.
8. To prevent enzymatic inactivation, all components in the kit should be stored at −20 °C; the reagent thawing process should be carried out on ice to avoid repeated freezing and thawing; the entire operation should be conducted on ice.
9. The number of cycles can be optimized depending on the type of cells, with a recommended cycle count of 14 for 100 pg of gDNA. For mammalian cells sorted by flow cytometry, it is suggested to set 17 cycles, and for single chromosomes, a range of 19–21 cycles is recommended.
10. It is important to ensure that the solution is thoroughly mixed and appears homogeneous. Insufficient mixing may compromise recovery and result in size selection.
11. Beads have been sufficiently dried when cracks just begin to emerge on the surface of the bead pellet and no liquid is visible in the tube. Overdrying the beads may result in dramatic yield loss. Overdrying is indicated by a dry, cracked appearance on the surface of the bead pellet.
12. When using the Agilent 2100 Bioanalyzer for WGA quality control, it is essential to ensure that the instrument is properly calibrated and that reagents are of high quality. Poor resolution or signal-to-noise ratio can lead to inaccurate sizing and misinterpretation of the electropherograms.
13. Previous studies have reported an impressive 100% concordance rate between the method for identifying tailings exceeding 10,380 base pairs and 1× coverage breadth over 70% at 50× average depth [20]. However, it is important to note that the relationship between tailing size and coverage breadth is not linear, and the method may require further refinement for different WGA applications.
14. In terms of troubleshooting, if low genome coverage breadth is observed, it may indicate insufficient DNA input or issues during the amplification process. Reevaluation of the WGA conditions or the use of additional quality control measures may be necessary.
15. For high-throughput applications, the tailing method can be adapted for use with other instruments such as the Agilent TapeStation 4200 or Advanced Analytical Fragment Analyzer, providing a more scalable solution for WGA quality control.

16. This tailing method has been specifically designed and validated for use with MALBAC products and is not suitable for application to other whole genome amplification (WGA) techniques.

Acknowledgments

This work was supported by the National Natural Science Foundation of China (grant no. 81372432) and other funding sources.

Conflict of Interest Statement The authors have no conflicts of interest to declare that are relevant to the content of this chapter.

References

1. Gawad C, Koh W, Quake SR (2016) Single-cell genome sequencing: current state of the science. Nat Rev Genet 17(3):175–188
2. Huang L, Ma F, Chapman A, Lu S, Xie XS (2015) Single-cell whole-genome amplification and sequencing: methodology and applications. Annu Rev Genomics Hum Genet 16: 79–102
3. Joosse SA, Gorges TM, Pantel K (2015) Biology, detection, and clinical implications of circulating tumor cells. EMBO Mol Med 7(1): 1–11
4. Navin N, Kendall J, Troge J, Andrews P, Rodgers L, McIndoo J et al (2011) Tumour evolution inferred by single-cell sequencing. Nature 472(7341):90–94
5. Ning L, Liu G, Li G, Hou Y, Tong Y, He J (2014) Current challenges in the bioinformatics of single cell genomics. Front Oncol 4:7
6. Sherman MA, Barton AR, Lodato MA, Vitzthum C, Coulter ME, Walsh CA et al (2018) PaSD-qc: quality control for single cell whole-genome sequencing data using power spectral density estimation. Nucleic Acids Res 46(4):e20
7. Hwang B, Lee JH, Bang D (2018) Single-cell RNA sequencing technologies and bioinformatics pipelines. Exp Mol Med 50(8):1–14
8. Hou Y, Wu K, Shi X, Li F, Song L, Wu H et al (2015) Comparison of variations detection between whole-genome amplification methods used in single-cell resequencing. Gigascience 4: 37
9. Leung ML, Wang Y, Waters J, Navin NE (2015) SNES: single nucleus exome sequencing. Genome Biol 16(1):55
10. Zong C, Lu S, Chapman AR, Xie XS (2012) Genome-wide detection of single-nucleotide and copy-number variations of a single human cell. Science 338(6114):1622–1626
11. Zhou X, Xu Y, Zhu L, Su Z, Han X, Zhang Z et al (2020) Comparison of Multiple Displacement Amplification (MDA) and Multiple Annealing and Looping-Based Amplification Cycles (MALBAC) in limited DNA sequencing based on tube and droplet. Micromachines (Basel). 11(7):645
12. Chen M, Song P, Zou D, Hu X, Zhao S, Gao S et al (2015) Correction: comparison of multiple displacement amplification (MDA) and multiple annealing and looping-based amplification cycles (MALBAC) in single-cell sequencing. PLoS One 10(4):e0124990
13. Chen M, Song P, Zou D, Hu X, Zhao S, Gao S et al (2014) Comparison of multiple displacement amplification (MDA) and multiple annealing and looping-based amplification cycles (MALBAC) in single-cell sequencing. PLoS One 9(12):e114520
14. Li N, Wang L, Wang H, Ma M, Wang X, Li Y et al (2015) The performance of whole genome amplification methods and next-generation sequencing for pre-implantation genetic diagnosis of chromosomal abnormalities. J Genet Genomics 42(4):151–159
15. Ning L, Li Z, Wang G, Hu W, Hou Q, Tong Y, et al. Quantitative assessment of single-cell whole genome amplification methods for detecting copy number variation using hippocampal neurons. Sci Rep 2015;5:11415
16. Stoecklein NH, Erbersdobler A, Schmidt-Kittler O, Diebold J, Schardt JA, Izbicki JR et al (2002) SCOMP is superior to degenerated

oligonucleotide primed-polymerase chain reaction for global amplification of minute amounts of DNA from microdissected archival tissue samples. Am J Pathol 161(1):43–51

17. Wong SQ, Li J, Tan AY, Vedururu R, Pang JM, Do H et al (2014) Sequence artefacts in a prospective series of formalin-fixed tumours tested for mutations in hotspot regions by massively parallel sequencing. BMC Med Genet 7: 23
18. Wang Y, Waters J, Leung ML, Unruh A, Roh W, Shi X et al (2014) Clonal evolution in breast cancer revealed by single nucleus genome sequencing. Nature 512(7513): 155–160
19. Sho S, Court CM, Winograd P, Lee S, Hou S, Graeber TG et al (2017) Precision oncology using a limited number of cells: optimization of whole genome amplification products for sequencing applications. BMC Cancer 17(1): 457
20. Chen C, Li J, Wan J, Lu Y, Zhang Z, Xu Z (2019) A low cost and input tailing method of quality control on multiple annealing, and looping-based amplification cycles-based whole-genome amplification products. J Clin Lab Anal 33(3):e22697

Chapter 4

Micromanipulation and Multiple-Strand Displacement Amplification for Forensic DNA Profiling of Single Sperm Cells

Glenn M. G. Theunissen and Richard Jäger

Abstract

In the clarification of sexual assault cases, forensic DNA profiling of single sperm cells can be desirable if only a few sperm cells are present in intimate swabs analyzed, or if sperm cells from more than one individual are present. However, the DNA content of a single sperm cell is far below the sensitivity limits of current standard forensic methods and thus does not allow for reliable DNA profiling. Here, we describe a protocol for micromanipulation of single sperm cells that is based on using an adhesive-coated tungsten needle to pick individual cells that have been spread on an agarose plate. Forensic DNA typing is thereafter accomplished by subjecting the extracted DNA to multiple strand displacement amplification (MDA) prior to analysis. For MDA, the QIAGEN REPLI-g Single Cell kit is used, and DNA profiles are analyzed using the Promega PowerPlex ESX 17 kit followed by capillary electrophoresis. Using this protocol, over 80% complete haploid DNA profiles can be obtained from the majority of single sperm cells picked.

Key words Forensic, DNA profile, Single sperm cells, Short tandem repeat, Whole genome amplification, Multiple strand displacement amplification, Low template DNA, Micromanipulation

1 Introduction

In the clarification of rape cases, intimate swabs are analyzed that typically contain a mixture of cells of more than one person. In cases with a single rapist, sperm cells of the rapist are mixed with vaginal (or anal epithelial) cells of the victim. The resulting mixed DNA profiles often have lower statistical power in terms of identification and can be difficult to interpret, and sometimes the perpetrator-derived component is obscured by the overwhelming amount of victim DNA [1].

To overcome these difficulties, procedures have been developed to enrich the sperm fraction [1, 2]. A challenge is posed, however, by cases where only few sperm cells, or sperm cells of more than one person (e.g., from the rapist and a regular sexual

Yafeng Ma (ed.), *Whole Genome Amplification: Methods and Protocols*, Methods in Molecular Biology, vol. 3003, https://doi.org/10.1007/978-1-0716-5060-8_4,

partner, or from several rapists in multiple perpetrator rape cases) are present. In such scenarios, the micromanipulation of single sperm cells from intimate swabs may be helpful, as this allows for the forensic DNA analysis of a pure sperm cell fraction or even of individual sperm cells [3–9].

Currently, DNA profiling is based on genotyping 15–20 standardized short tandem repeat (STR) loci using multiplex PCR, followed by capillary electrophoresis [10]. The PCR assays used typically require at least 60 pg of DNA to yield complete DNA profiles [11–14]. A single sperm cell contains approximately 3.3 pg of DNA, and thus, forensic STR analysis will at best yield partial haploid profiles that display variability due to stochastic effects [8]. Moreover, complete haploid profiles of at least 9 sperm cells need to be combined in order to compile a diploid donor profile with 95% statistical confidence [15].

One idea to improve the analytical sensitivity of multiplex PCR-based STR analysis is to increase the number of template DNA copies beforehand using whole genome amplification (WGA) [16, 17]. In forensic DNA analysis, the application of WGA has so far been of limited success, possibly because of the low quality of trace DNA, coextracted impurities, the proneness of STR loci to amplification-based artifacts, and the fundamental inability to overcome stochastic sampling effects (reviewed in [10]).

In recent work, however, we have shown that it is possible to obtain over 80% complete haploid STR profiles of individual sperm cells if subjecting their genomic DNA to multiple strand displacement amplification (MDA)-based WGA before standard STR analysis [9]. MDA is based on random-primer-initiated replication of template DNA that has been rendered single-stranded [18]. The phi 29 polymerase used has strand displacement activity, thus generating novel single-stranded template DNA once reaching a preceding, newly generated double strand [19]. By this means, amplification is quasi-exponential and can be carried out isothermally. The method requires long stretches of intact DNA as can be extracted from sperm cells [20]. Due to its high fidelity and processivity the phi 29 polymerase can generate amplification products of over 20 kb in length [19] with only a few amplification artifacts at STR loci [21].

In this chapter, we describe our protocol for micromanipulation of individual sperm cells using a tungsten needle that has been coated with a water-soluble adhesive [22] to which sperm cells adhere. After picking, the genomic DNA is released from the captured sperm cell and subjected to WGA using the MDA-based REPLI-g kit. Following cleanup, the WGA products are subjected to multiplex-PCR-based analysis of 16 STR loci (plus the sex marker amelogenin) using the Promega PowerPlex ESX 17 kit, followed by capillary electrophoretic analysis.

2 Materials

2.1 Releasing Spermatozoa from Cotton Swabs

1. Bleach solution, e.g., "Sagrotan Schimmelfrei" (Reckitt Benckiser Deutschland GmbH, Heidelberg, Germany) (*see* **Note 1**).
2. 70% ethanol solution.
3. Cotton mock intimate swabs with spermatozoa (frozen or nonfrozen) (*see* **Note 2**).
4. Thermoshaker.
5. Scalpel.
6. Glass petri dish.
7. Tweezers.
8. 1.5 mL microcentrifuge tubes.
9. Buffer ATL (QIAGEN, Hilden, Germany).
10. Proteinase K (QIAGEN, Hilden, Germany).
11. Vortexer.
12. Centrifuge for 1,5 mL centrifuge tubes.
13. MiniSpin Centrifuge (optional).
14. pluriStrainer Mini 10 μm cell strainer (pluriSelect Life Science, Leipzig, Germany).
15. Phosphate-buffered saline (PBS).
16. DNA-free water.
17. Microscope slides.
18. Light microscope with 200× magnification.

2.2 Spreading of Sperm Cells Onto Agarose Plates

1. Microwave oven.
2. 90 mm plastic petri dishes.
3. Agarose (electrophoresis grade).
4. DNA-free water.

2.3 Preparation of Tungsten Needle and Micromanipulation

1. Inverted microscope. e.g., Axiovert S 100 (Zeiss, Oberkochen, Germany) equipped with a micromanipulator, e.g., Transfer-Man micromanipulation system (Eppendorf, Hamburg, Germany).
2. 3 M double-sided water-soluble adhesive tape (3 M, USA).
3. Tungsten needle with a tip diameter of 5 μm (PLANO GmbH, Wetzlar, Germany).
4. 0,2 mL PCR tubes.
5. DNA-free water.

2.4 Whole Genome Amplification

1. REPLI-g Single Cell WGA kit (QIAGEN, Hilden, Germany).
2. innuPREP PCRpure Kit (Analytik Jena, Jena, Germany).
3. Two heating blocks.

2.5 STR Analysis

1. PowerPlex ESX 17 kit (Promega, Fitchburg, WI, USA).
2. Thermal cycler with heated lid.
3. Deionized HiDi™ formamide (Thermo Fisher, Waltham, USA).
4. ABI Prism 310 or 3130 Genetic Analyzer with GeneMapper™ software (Thermo Fisher, Waltham, USA).

3 Methods (*See* Note 3)

3.1 Releasing Spermatozoa from Cotton Swabs

If using a cell suspension of spermatozoa instead of swabs, **steps 1–20** can be skipped.

1. Thaw swabs at room temperature for 30 min when using swabs that have been stored frozen.
2. Turn on a thermoshaker and set the temperature at 56 °C.
3. On a DNA-free glass petri dish, cut the cotton part off from the swab's end using a scalpel and split the cotton part vertically in half.
4. Transfer each cotton half into a 1.5 mL microcentrifuge tube using a pair of tweezers.
5. Add 500 μL of buffer ATL and 25 μL of Proteinase K to each tube (*see* **Note 4**). Vortex the tubes for 10 s in order to unfold the cotton.
6. Incubate the tubes in a thermoshaker at 56 °C with 900 rpm shaking for 30 min.
7. Vortex the tubes again for 10 s.
8. Incubate for another 30 min at 56 °C with 900 rpm shaking.
9. Vortex the tubes for 10 s and briefly spin down.
10. Place each cotton part into a pluriStrainer Mini 10 μm cell strainer (pluriSelect Life Science, Leipzig, Germany) with 10 μm pore size (*see* **Note 5**).
11. Put the cell strainer back into the tube and centrifuge at 14,000 g for 5 min to release the remaining liquid and spermatozoa left on the cotton parts.
12. Discard the cell strainer containing the cotton part and the supernatant without disturbing the pellet.
13. Add 500 μL of PBS to each tube.
14. Resuspend the pellet by tapping and pulse-vortexing (*see* **Note 6**).

15. Centrifuge at 14,000 g for 5 min.
16. Discard the supernatant without disturbing the pellet (*see* **Note 7**).
17. Resuspend the pellet in 300 μL of DNA-free water by tapping and pulse-vortexing (*see* **Note 8**).
18. Briefly spin down the tubes to remove any liquid on the top of the tube and lid.
19. Place 20 μL of the cell suspension on a microscope slide.
20. Analyze with a light microscope at 200× magnification (*see* **Note 9**).
21. If the suspension is highly concentrated with spermatozoa, dilute the suspension with DNA-free water. A cell suspension of not more than 100 spermatozoa per 100 μL should be aimed at to sufficiently separate the spermatozoa for the micromanipulation steps.

3.2 Spreading of Sperm Cells Onto Agarose Plates (See Note 10)

1. Suspend 0,25 g agarose in 25 mL deionized DNA-free water to make a 1% (w/v) agarose solution in a 100 mL Erlenmeyer.
2. Dissolve the agarose by boiling in a microwave oven shortly for three times or until dissolved completely (*see* **Note 11**).
3. Pour the melted agarose into 90 mm petri dishes to create a thin layer of approximately 3 mm thickness.
4. Allow the agarose to solidify by cooling down to room temperature with the petri dish covered with a slightly lifted lid to prevent airborne contamination while allowing for drying.
5. Let the agarose dry for 30 min.
6. Spread 100–300 μL of the cell suspension onto the agarose gel petri dishes, aiming for 50–100 sperm cells per dish (*see* **Note 12**).
7. Place the agarose gel petri dish containing sperm cells under an inverted microscope (e.g., Axiovert S 100, Zeiss, Oberkochen, Germany) equipped with a micromanipulation system (e.g., TransferMan, Eppendorf, Hamburg, Germany), as shown in Fig. 1.
8. Search for spermatozoa at 200× magnification.

3.3 Preparation of Tungsten Needle and Micromanipulation

1. Collect a small amount of adhesive from a piece of 3 M double-sided water-soluble adhesive tape (3 M, USA) using a tungsten needle with a tip diameter of 5 μm (*see* Fig. 2 and **Notes 13** and **14**).
2. Place the tungsten needle with adhesive onto the micromanipulation system (see Fig. 1).

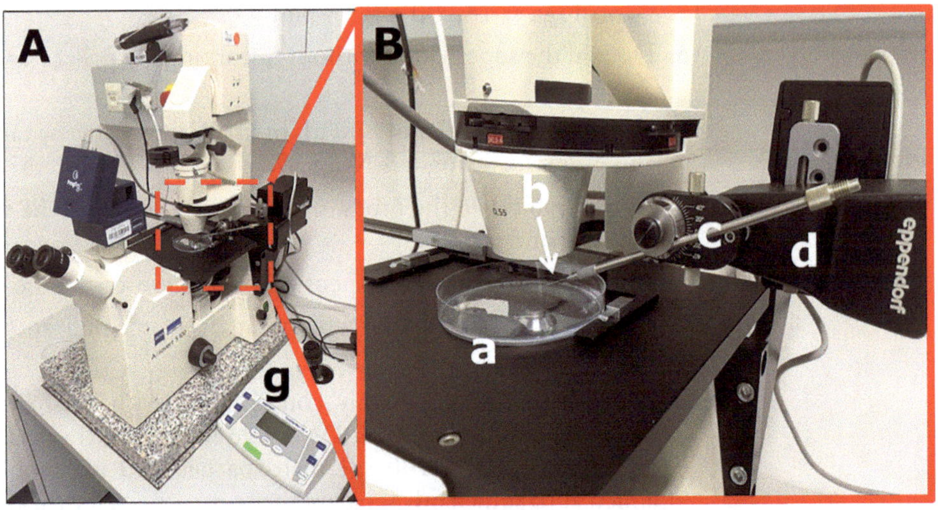

Fig. 1 Needle and agarose gel-based micromanipulation set-up. (**A**) Fully equipped inverted microscope with additional joystick (g) to control the needle movements. (**B**) Close up of the agarose filled petri dish (a) ready to perform micromanipulation of spermatozoa by using a tungsten needle (b) inserted into the needle holder (c) that is moved by the arm (d)

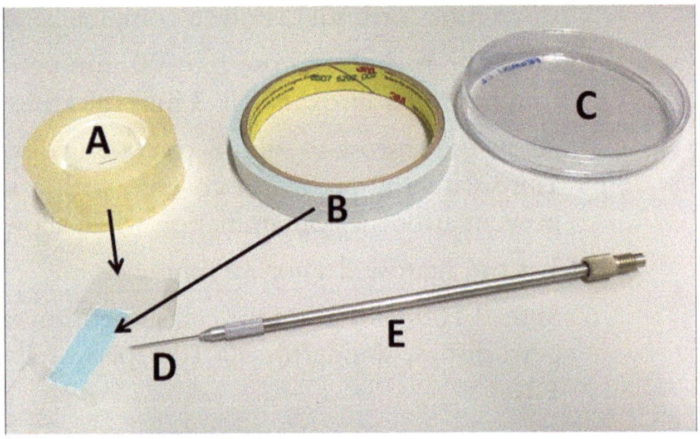

Fig. 2 Components used for tungsten needle-based micromanipulation of spermatozoa. A piece of the 3 M water-soluble adhesive tape (B) is fastened to a clean work surface with multipurpose transparent adhesive tape (A). A tungsten needle (D) is inserted in a capillary/needle holder (E). Adding 3 M soluble adhesive onto the needle (D) is accomplished by scraping the tip of the needle on the piece of fastened 3 M tape. An agarose gel-filled petri dish (C) serves for spreading out the sperm cell suspension for micromanipluation

3. Using the adhesive on the needle tip, pick a single sperm cell from the agarose gel surface. The typical appearance of the adhesive zone touching the agarose on top of a single sperm cell is shown in Fig. 3 (*see* **Note 15**).
4. Transfer the sperm cell into a PCR tube containing 4 μL of DNA-free water by swirling the needle tip in the water for 10 s to dissolve the adhesive and release the spermatozoa into the liquid (*see* **Note 16**).

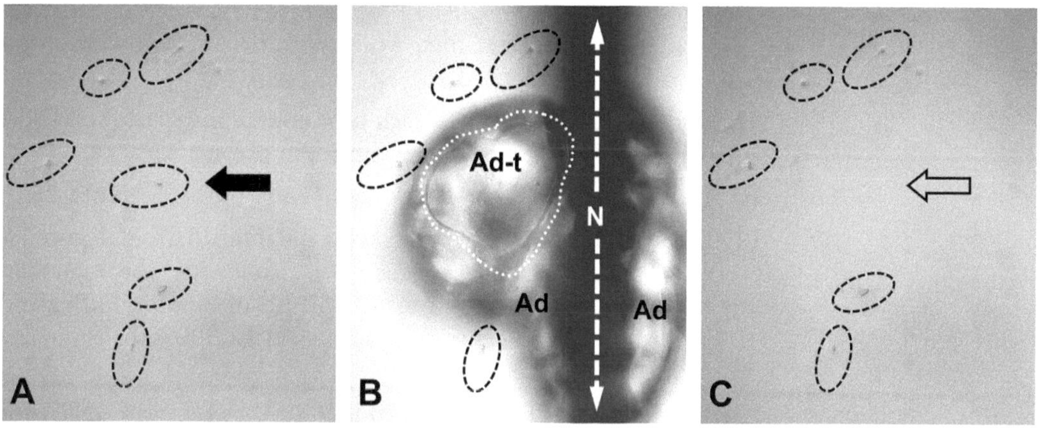

Fig. 3 Microscopic view of the isolation of a single sperm cell as seen at 200x magnification. (**a**) Before micromanipulation. Sperm cells are surrounded by dashed black lines; the sperm cell to be picked is marked by a black arrow. (**b**) Micromanipulation using the tungsten needle tip (N, seen as a black shadow and indicated by the dashed white arrows), coated with adhesive (Ad). The area where the adhesive touches the agarose (Ad-t) is visible by a fine demarcation line that is enclosed by the white dotted line on the figure. (**c**) After picking. The open arrow points to the empty region from which the sperm cell has been picked

5. After each transfer, clean the needle tip with bleach solution, rinse with DNA-free water twice, and wipe dry with a clean lab tissue.

3.4 Whole Genome Amplification

Whole genome amplification is done using the REPLI-g Single Cell WGA kit, following the manufacturer's protocol "Amplification of Genomic DNA from Single Cells," though the incubation step at 30 °C is extended for 16 h (*see* **Note 17**). All components used in **steps 1–9** are provided by the kit.

1. Set one heating block at 65 °C, one at 30 °C.
2. Thaw the kit components at room temperature. The REPLI-g sc polymerase has to be thawed on ice (*see* **Note 18**).
3. Reconstitute buffer DLB by adding 500 μl H_2O to the tube provided by the kit. Mix thoroughly and centrifuge briefly to dissolve (*see* **Note 19**).
4. Prepare Buffer D2 by adding 3 μL of 1 M DTT to 33 μL of the reconstituted buffer DLB from **step 2** (*see* **Note 20**).
5. Add 3 μL of buffer D2 to the tubes containing the sperm cells. Mix carefully by flicking the tube and centrifuge briefly (*see* **Notes 21** and **22**).
6. Incubate for 10 min at 65 °C.
7. Add 3 μL Stop Solution. Mix carefully by flicking the tube and centrifuge briefly. Store on ice.
8. Prepare a master mix as follows: To 9 μL H_2O, add 29 μL REPLI-g sc Reaction Buffer, then briefly vortex and spin down.

Thereafter, add 2 μL REPLI-g sc DNA Polymerase to finally yield a 40 μL master mix, which is sufficient for one reaction (*see* **Notes 23** and **24**).

9. Add 40 μL master mix to each tube containing sperm cell DNA (from **step 6**), briefly vortex, and spin down.
10. Incubate at 30 °C for 16 h on a heating block (*see* **Note 25**).
11. Clean up the WGA products with the innuPREP PCRpure kit (Analytik Jena, Jena, Germany) by following the manufacturer's protocol, though using 450 μL of binding buffer and repeating the first centrifugation step at 12,000 g for 2 min (*see* **Note 26**).
12. Dilute the purified WGA product 1:50 as template DNA for the subsequent STR analysis and store at 4 °C (*see* **Note 27**).

3.5 STR Analysis

1. Set up a forensic autosomal multiplex PCR with a forensic multiplex PCR kit, e.g., the PowerPlex ESX 17 kit in a pre-PCR room (*see* **Note 28**).
2. Add 1 μL of 1:50 diluted purified WGA product as template DNA to a 4 μL PCR amplification mix (containing 1 μL of PowerPlex® ESX 5X Master Mix, 0.5 μL of PowerPlex® ESX 17 10X Primer Pair Mix, 2.5 μL water) in a 0,2 mL PCR tube (*see* **Note 29**).
3. Amplify in a thermal cycler with heated lid by following the PCR amplification program with 1 cycle of 94 °C for 4 min, 5 cycles of 96 °C for 30 s, 63 °C for 120 s, and 72 °C for 75 s, followed by 25 cycles of 94 °C for 30 s, 60 °C for 120 s, and 72 °C for 75 s, and one final extension at 68 °C for 60 min.
4. Analyze PCR products with capillary electrophoresis on, e.g., an ABI Prism 310 or 3130 Genetic Analyzer by first denaturing 1 μL of PCR product in 12 μL deionized HiDi™ formamide and 0,5 μl WEN ILS 500 size standard (from PowerPlex ESX 17 kit) at 95 °C for 4 min.
5. Inject the denatured samples at 3 kV for 3 s (*see* **Note 30**).
6. Genotype the generated data with the GeneMapper™ software version by setting the allele calling threshold at 50 RFU's or at your laboratory's preferred analytical threshold (*see* **Notes 31–34**).

4 Notes

1. To prevent DNA contamination, all work surfaces should be treated with a bleach-based solution such as "Sagrotan Schimmelfrei" (Reckitt Benckiser Deutschland GmbH, Germany) and rinsed with 70% ethanol. Pre-PCR tools such as pipettes,

pencils, tweezers, tube racks, vortexers, centrifuges, and microscopes should also be wiped with bleach solution and rinsed with 70% ethanol.

2. Intimate mock swabs can be prepared by swabbing the inside of a person's cheeks to collect buccal epithelial cells, followed by an incubation at room temperature until dried (best overnight). The swab is then spiked with 100 ml of 1:50 diluted semen (diluted in PBS) and dried at room temperature again. Intimate (mock) swabs can be stored at room temperature or at 4 °C, but it's not recommended to freeze them to prevent cellular damage.
3. Due to the enormous sensitivity of the WGA and PCR, it is crucial to take measures to prevent contamination with unrelated DNA. Therefore, pre-amplification work (including micromanipulation) should be performed in a pre-PCR room, with personnel wearing hair nets and mouth masks. All mastermixes for PCR and WGA should be prepared inside a PCR UV-cabinet. Negative controls of the reagents should be included, and the DNA profile of each person conducting the experiment should be generated to check for any contamination incidents.
4. **Steps 5–10** serve for releasing the spermatozoa from the cotton tip and at the same time for lysing the non-sperm cells.
5. The cell strainer serves to separate the sperm cells from larger particles, debris, and cotton fibers.
6. A pellet might not always be visible. When a pellet is visible, we apply short, gentle pulses of approximately 1 s in approximately 2 s intervals until we see it dissolved. This typically requires 5 pulses, so we also use five pulses when no pellet is visible. We do not resuspend with a pipette because sperm cells might remain attached to the pipette tip.
7. We use a micropipette with 200 μL pipette tips, with slowly pipetting and avoiding to touch the area of the pellet (or the presumed area, if the pellet is not visible). Leave some liquid (few μL) in the tube to avoid removing the pellet.
8. Pulse-vortexing as in **Note 6**. In this step, the PBS is replaced by water in order to avoid crystal formation by the drying solvent after spreading the cells on an agar plate for micromanipulation.
9. Be aware of the lack of the spermatozoa's tail. Spermatozoa often lose their tails when dried on a swab. Of most spermatozoa, only the heads will be present.
10. During micromanipulation work, the operator should wear gloves and a face mask, and all pipetting steps before and after micromanipulation should be performed in a PCR UV-cabinet.

11. Wear safety glasses and use heat-resistant gloves when handling the flask containing hot agarose. Please note that boiling delay may occur when taking out the heated agarose from the microwave oven.
12. This is best done by gently swirling the dish to spread the suspension evenly over the majority of the agarose surface.
13. The adhesive tape should be fixed to a surface, with the adhesive at the top side, as shown in Fig. 2, and the tungsten needle tip should be gently scraped along the adhesive.
14. If trying out other brands of adhesive tape, it is important to test whether the adhesive is compatible with the subsequent WGA and PCR steps. To this end, a mock assay should be performed with genomic DNA mixed with 4 μL DNA-free water into which the adhesive-coated tungsten needle tip is placed, followed by WGA and PCR steps as described in Subheadings 3.4 and 3.5.
15. In most cases, the absence of the sperm cell from the touched area after removal of the tungsten needle indicates that the cell is attached to the adhesive. However, the attachment cannot be controlled, and cells not remaining attached may be the cause of some samples not yielding STR profiles.
16. Dissolution of the adhesive indicates release of the sperm cell. Check if the adhesive completely dissolved from the needle by placing the needle back on the micromanipulation system. No adhesive residues should be seen on the tip of the needle.
17. Sperm cell lysis and release of DNA require DTT, which is contained in the D2 buffer. The cell lysis principle is an alkaline lysis which at the same time denatures the genomic DNA. The addition of the stop solution serves to restore a pH that is optimal for the amplification reaction. The amplification is carried out by the addition of REPLI-g sc Reaction buffer and REPLI-g DNA polymerase (which is phi 29 polymerase).
18. All buffers and reagents should be vortexed and briefly spun down before use to ensure homogeneity. The REPLI-g sc Reaction Buffer may form a precipitate after thawing, which will dissolve by vortexing for 10 s.
19. Reconstituted buffer DLB can be stored for 6 months at −20 °C.
20. The total volume of 36 μL is sufficient for 12 reactions. If more or less reactions are required, the volumes may be scaled up or down accordingly.
21. Due to the dissolved adhesive, the final volume of the sperm cell suspension may be slightly above the 4 μL recommended by the REPLI-g protocol. We nevertheless strictly followed the REPLI-g protocol by adding 3 μL of lysis buffer D2.

22. Do not mix by pipetting to avoid sticking of cells to the pipette tip.
23. Vortexing is not recommended by the manual. Proper mixing is presumably achieved anyway by the subsequent pipetting of the 40 μL into the DNA tubes. We nevertheless gently stirred with the pipette tip while adding the polymerase. The master mix should be kept on ice and used immediately upon addition of REPLI-g sc DNA Polymerase.
24. If a master mix for multiple reactions is prepared, this mixture should be scaled up according to the number of reactions plus 10% (e.g., for 10 reactions, a mix for 11 reactions is to be prepared). In this case, however, the master mix for the multiple reactions should again be briefly vortexed and spun down after the addition of the polymerase to ensure a homogenous mixture and hence a homogenous distribution of the polymerase to all samples.
25. If a thermal cycler with a heated lid is used, the temperature of the lid should be set to 70 °C.
26. The innuPREP PCRpure kit is suited for 50 μL of PCR reactions. Although the final volume of the REPLI-g reaction was only 40 μL, we experienced difficulties fitting this volume onto the columns when using 500 μL binding buffer. Reducing the volume of binding buffer to 450 μL solved the problem.
27. If required, the DNA concentration of the 1:50 WGA product can be quantified by human DNA-specific qPCR. Please note that the REPLI-g single cell assay is known to generate template-independent products, and thus, a nonhuman specific DNA quantitation will be misleading. For autosomal STR typing, we did not quantitate the DNA. If on the electropherograms STR loci dropped out, we, in addition, used the undiluted WGA products for STR typing; if signals were too strong, we used a 1:100 dilution.
28. Any STR kit might work—we give a quick overview of our procedure using the PowerPlex ESX 17 kit. It is important to follow the instructions of the manufacturer of both the STR kit and the capillary electrophoresis device. The researcher is expected to be an expert with this kind of analysis; thus, we only point out the steps where we find some additional notes necessary.
29. The manufacturer's instructions for the PowerPlex ESX 17 kit recommend a total reaction volume of 25 μL for the multiplex PCR; however, we successfully use a 5 μL reaction volume.
30. This setting will depend on the type of genetic analyzer used. We recommend following the manufacturer's advised settings.

31. Since the MDA reaction often leads to pronounced inter-loci imbalances, allele peaks with extremely high RFUs (off-scale peaks) on electropherograms can appear and can blead into other color detection channels. This complicates allele calling. It is therefore important to switch on the off-scale peak indicator in GeneMapper™ to indicate overamplified peaks.
32. Usually, overamplified PCR products are diluted prior to fragment analysis in conventional forensic STR analysis to reduce peak heights and prevent off-scale peaks. Though by strongly diluting PCR products to improve overamplified loci, low allele peaks can disappear (locus dropout) in loci that are underamplified. To solve this issue, it is strongly advised to run fragment analysis first undiluted to detect underamplified alleles, and then dilute the PCR product, followed by running a second fragment analysis to improve allele calling in the presence of overamplified alleles.
33. As a haploid STR profile is expected, only one allele peak should be seen in each autosomal STR locus. More than one peak at STR loci is thus indicative of DNA contamination.
34. Depending on the forensic STR kit and the PCR reaction volume used, the amount of template DNA needed can differ, and it is therefore advised to test what amount of template DNA is appropriate for the STR kit used.

Acknowledgments

Development of this protocol was funded by the German Federal Ministry of Education and Research (project no. 03FH019PX4) and by a PhD scholarship to GMGT from the Institute of Safety and Security Research (ISF) of Bonn-Rhein-Sieg University of Applied Sciences.

References

1. Chapman BR, Blackwell SJ, Muller LH (2020) Forensic techniques for the isolation of spermatozoa from sexual assault samples – a review. Forensic Sci Rev 32(2):105–116
2. Gill P, Jeffreys AJ, Werrett DJ (1985) Forensic application of DNA 'fingerprints'. Nature 318(6046):577–579. https://doi.org/10.1038/318577a0
3. Auka N, Valle M, Cox BD, Wilkerson PD, Dawson Cruz T, Reiner JE et al (2019) Optical tweezers as an effective tool for spermatozoa isolation from mixed forensic samples. PLoS One 14(2):e0211810. https://doi.org/10.1371/journal.pone.0211810
4. Elliott K, Hill DS, Lambert C, Burroughes TR, Gill P (2003) Use of laser microdissection greatly improves the recovery of DNA from sperm on microscope slides. Forensic Sci Int 137(1):28–36. https://doi.org/10.1016/s0379-0738(03)00267-6
5. Li CX, Han JP, Ren WY, Ji AQ, Xu XL, Hu L (2011) DNA profiling of spermatozoa by laser capture microdissection and low volume-PCR. PLoS One 6(8):e22316. https://doi.org/10.1371/journal.pone.0022316
6. Ostojic L, O'Connor C, Wurmbach E (2021) Micromanipulation of single cells and fingerprints for forensic identification. Forensic Sci

Int Genet 51:102430. https://doi.org/10.1016/j.fsigen.2020.102430

7. Pereira J, Neves R, Forat S, Huckenbeck W, Olek K (2012) MtDNA typing of single-sperm cells isolated by micromanipulation. Forensic Sci Int Genet 6(2):228–235. https://doi.org/10.1016/j.fsigen.2011.05.005
8. Schneider C, Muller U, Kilper R, Siebertz B (2012) Low copy number DNA profiling from isolated sperm using the aureka(R)-micromanipulation system. Forensic Sci Int Genet 6(4):461–465. https://doi.org/10.1016/j.fsigen.2011.10.001
9. Theunissen GMG, Gibb A, Lin Paul KT, Rolf B, Forat S, Jäger R (2021) DNA profiling of single sperm cells after whole genome amplification. Forensic Sci Int Rep 4. https://doi.org/10.1016/j.fsir.2021.100240
10. Jager R (2022) New perspectives for whole genome amplification in forensic STR analysis. Int J Mol Sci 23(13). https://doi.org/10.3390/ijms23137090
11. Ensenberger MG, Lenz KA, Matthies LK, Hadinoto GM, Schienman JE, Przech AJ et al (2016) Developmental validation of the PowerPlex((R)) fusion 6C system. Forensic Sci Int Genet 21:134–144. https://doi.org/10.1016/j.fsigen.2015.12.011
12. Kraemer M, Prochnow A, Bussmann M, Scherer M, Peist R, Steffen C (2017) Developmental validation of QIAGEN investigator ((R)) 24plex QS kit and investigator((R)) 24plex GO! Kit: two 6-dye multiplex assays for the extended CODIS core loci. Forensic Sci Int Genet 29:9–20. https://doi.org/10.1016/j.fsigen.2017.03.012
13. Martin P, de Simon LF, Luque G, Farfan MJ, Alonso A (2014) Improving DNA data exchange: validation studies on a single 6 dye STR kit with 24 loci. Forensic Sci Int Genet 13:68–78. https://doi.org/10.1016/j.fsigen.2014.07.002
14. RS ML, Bourdeau-Heller J, Patel J, Thompson JM, Pagram J, Loake T et al (2014) Developmental validation of the PowerPlex((R)) ESI 16/17 fast and PowerPlex((R)) ESX 16/17 fast systems. Forensic Sci Int Genet 13:195–205. https://doi.org/10.1016/j.fsigen.2014.08.004
15. Lucy D, Curran JM, Pirie AA, Gill P (2007) The probability of achieving full allelic representation for LCN-STR profiling of haploid cells. Sci Justice 47(4):168–171. https://doi.org/10.1016/j.scijus.2007.10.001
16. Ballantyne KN, van Oorschot RA, Mitchell RJ (2007) Comparison of two whole genome amplification methods for STR genotyping of LCN and degraded DNA samples. Forensic Sci Int 166(1):35–41. https://doi.org/10.1016/j.forsciint.2006.03.022
17. Barber AL, Foran DR (2006) The utility of whole genome amplification for typing compromised forensic samples. J Forensic Sci 51(6):1344–1349. https://doi.org/10.1111/j.1556-4029.2006.00262.x
18. Czyz ZT, Kirsch S, Polzer B (2015) Principles of whole-genome amplification. Methods Mol Biol 1347:1–14. https://doi.org/10.1007/978-1-4939-2990-0_1
19. Blanco L, Bernad A, Lazaro JM, Martin G, Garmendia C, Salas M (1989) Highly efficient DNA synthesis by the phage phi 29 DNA polymerase. Symmetrical mode of DNA replication. J Biol Chem 264(15):8935–8940
20. Giusti A, Baird M, Pasquale S, Balazs I, Glassberg J (1986) Application of deoxyribonucleic acid (DNA) polymorphisms to the analysis of DNA recovered from sperm. J Forensic Sci 31(2):409–417
21. Canceill D, Viguera E, Ehrlich SD (1999) Replication slippage of different DNA polymerases is inversely related to their strand displacement efficiency. J Biol Chem 274(39):27481–27490. https://doi.org/10.1074/jbc.274.39.27481
22. Farash K, Hanson EK, Ballantyne J (2015) Enhanced genetic analysis of single human bioparticles recovered by simplified micromanipulation from forensic 'touch DNA' evidence. J Vis Exp (97):52612. https://doi.org/10.3791/52612

Chapter 5

Whole-Genome Amplification of Single Circulating Tumor Cells from Mice Xenografts

Nuria Estévez-Gómez, Cristóbal Fernández-Santiago, Roberto Piñeiro, and David Posada

Abstract

The genomic analysis of circulating tumor cells (CTCs) can help us identify their specific characteristics and reveal intratumor heterogeneity while also reducing the need for invasive sampling. However, studying single CTCs can be challenging, as whole-genome amplification (WGA) is needed to obtain enough genetic material for high-throughput sequencing. Here, we present a protocol for isolating single CTCs from mouse xenografts, followed by WGA using the Ampli1WGA Plus kit.

Key words Circulating tumor cells, Liquid biopsy, Single-cell whole-genome amplification, Single-cell genomics, Cancer genomics, Ampli1, Xenografts

1 Introduction

Single-cell genomic studies typically require a whole-genome amplification (WGA) step, which introduces technical biases that can compromise the interpretation of single-cell DNA sequencing data [1]. Several strategies have been developed for single-cell WGA (scWGA) based on PCR [2, 3], multiple displacement amplification (MDA) [4–6], or a combination of both [7, 8]. These strategies perform differently depending on the parameters of interest [9].

Circulating tumor cells (CTCs) are cancer cells that can detach from tumors and enter the bloodstream. However, they are considered rare populations (1 CTC: billions of blood cells), as most of them die in circulation, and only a small fraction is capable of colonizing other tissues [10, 11]. The study of CTCs enables the investigation of intratumoral heterogeneity using non-invasive procedures such as liquid biopsy [12–14].

Our protocol was developed for the study of CTCs from mouse breast cancer xenografts. First, blood samples are collected from the

Yafeng Ma (ed.), *Whole Genome Amplification: Methods and Protocols*, Methods in Molecular Biology, vol. 3003, https://doi.org/10.1007/978-1-0716-5060-8_5,

mice's hearts, and the CTCs are enriched using the microfluidic Parsortix® technology, which takes advantage of the larger size and lower deformability of CTCs compared to other blood cells. After the enrichment, single CTCs are isolated by manual micromanipulation. For WGA, an adaptor-ligation PCR-based scWGA method (Ampli1 WGA PLUS) is used. Following cell lysis, DNA is fragmented using the restriction enzyme *MseI* (3) at TTAA sequences, resulting in ~19 million DNA fragments [15] with sticky ends on both sides. A complementary adaptor is added, and after filling the ends, it is used as a priming site for a single universal PCR primer [16]. WGA products undergo a quality control (QC) step using a PCR-based QC kit to amplify four markers distributed across the human genome (chromosomes 5q, 6q, 12p, and 17p) with varying amplicon lengths. This QC assesses whether the WGA amplicons are evenly distributed across the genome. Samples positive for at least three markers are considered to have passed the QC. To avoid wasting sequencing resources, adaptors are removed using the *MseI* enzyme, followed by a purification step before preparing the sequencing library.

2 Materials

Prepare and store all reagents at room temperature (unless indicated otherwise). Diligently follow all waste disposal regulations when disposing of waste materials.

2.1 Blood Sample Collection

1. Isoflurane vaporizer with induction chamber and nose cone.
2. Electric shaver or fine scissors.
3. 1 mL syringe +25–27 G needle.
4. Ethylenediaminetetraacetic acid (EDTA) blood-collection tubes.
5. EDTA solution (e.g., EDTA-coated syringe or 10–20 μL of 0.5 M EDTA in syringe) to preload the syringe.

2.2 CTCs Enrichment

1. Parsortix PR1 instrument (Angle PLC).
2. Parsortix GEN3P6.5 separation cassettes (6.5 μm).
3. Parsortix GEN3P10 cleaning cassettes (10 μm).
4. Phosphate-buffered saline (PBS): 1X solution.
5. Alcohol wipes and lint-free cloths.
6. Paraformaldehyde (PFA): 4% solution in PBS.

2.3 Single CTC Isolation

1. Sterile chambered cell culture slide.
2. Borosilicate thin-wall capillaries, 1.0 mm OD, 0.78 mm ID, 75 mm.

3. Bovine serum albumin (BSA).
4. Fetal bovine serum (FBS).
5. Manual micromanipulator (Leica Microsystems).
6. CellTram Oil mechanical microinjector (Eppendorf).
7. Inverted fluorescence microscope.

2.4 Single-Cell WGA

1. Ampli1™ WGA Plus kit (Menarini Silicon Biosystems).
2. REPLI-g human control kit (QIAGEN): 10 ng of human gDNA.
3. Ampli1™ WGA QC kit (Menarini Silicon Biosystems).
4. NEBuffer 4 10X: 50 mM Potassium Acetate, 20 mM Tris-acetate, 10 mM Magnesium Acetate, 1 mM DTT.
5. MseI enzyme 50 U/μL.
6. AMPure XP beads (Beckman Coulter).
7. Magnetic stands for PCR tubes and plates.
8. Ethanol: 70% solution in water.
9. Nuclease-free water (NFW).
10. Qubit dsDNA BR assay kit (Thermo Fisher Scientific).
11. TapeStation D1000 assay kit (Agilent Technologies).
12. Qubit 4 fluorometer (Thermo Fisher Scientific).
13. 2200 TapeStation (Agilent Technologies).

3 Methods

3.1 Blood Extraction by Intracardiac Puncture from Tumor-Bearing Mice

The collection of blood from the mouse heart is a terminal procedure and must be performed under deep general anesthesia. Alternatively, it can be done on the euthanized mouse (*postmortem*).

3.1.1 Blood Collection Under General Anesthesia

1. Induce deep anesthesia by placing the mouse in the induction chamber. Administer 3–4% isoflurane in oxygen until the pedal reflex is lost (surgical plane) (*see* **Note 1**).
2. To maintain anesthesia, transfer the mouse to the nose cone, placing it dorsally. Reduce the isoflurane concentration to 2–3% for maintenance of anesthesia.
3. Prepare the chest for injection. Shave the left lateral thorax, just below the sternum, where the heart lies tilted to the left (*see* **Notes 2** and **3**).
4. Gently lift the skin and visualize the heartbeat (*see* **Note 4**).
5. It is advisable to coat the inside of the syringe with 0.5 M EDTA by drawing in 100–200 μL and then expelling it before blood collection to help prevent clotting during the procedure.

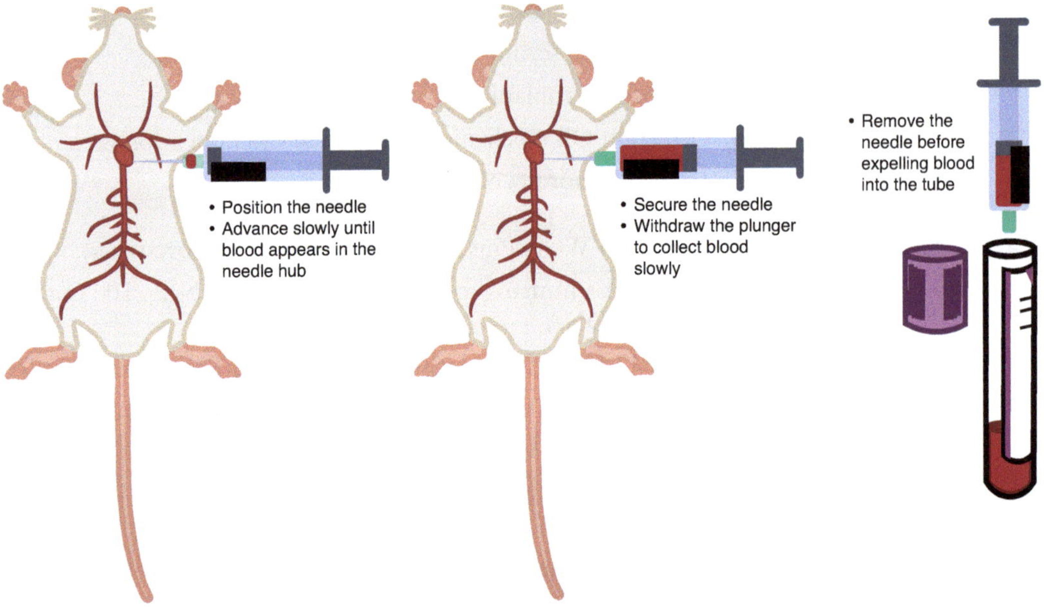

Fig. 1 Illustration of blood collection via cardiac puncture in a mouse. The procedure involves positioning the needle perpendicular to the chest wall and advancing it slowly until blood appears in the needle hub. Once the ventricle is accessed, the needle is secured, and the plunger is withdrawn gradually to collect blood. After collection, the needle is removed before the blood is expelled into an EDTA collection tube

6. Hold the syringe by the barrel, and with the needle bevel up, insert it perpendicular to the chest wall at the level of the left elbow (left ventricular apex) (*see* **Note 5**).
7. Penetrate only with the bevel tip at first. Apply light counter-pressure over the right side of the animal with your non-dominant hand, then advance slowly until blood appears in the needle hub (blood flashback) (Fig. 1).
8. Secure the needle to the syringe and gently withdraw the plunger to collect blood slowly, minimizing hemolysis.
9. Confirm the ventricular entry by the color of the blood. A bright red color means oxygenated blood from the left ventricle, while a brownish-red color means deoxygenated blood from the right ventricle. You may advance slightly for right-ventricular sampling if desired (*see* **Note 6**).
10. Continue gentle aspiration until the flow stops or the syringe is full (≈0.8–0.9 mL).
11. After blood flow stops, remove the needle before expelling blood into the collection tube to prevent cell lysis (Fig. 1; *see* **Notes 7** and **8**).
12. Immediately after collecting the blood sample, perform cervical dislocation on the mouse to complete the procedure.

3.2 CTC Enrichment with Parsortix®

The Parsortix® system (Fig. 2) protocol comprises three major procedures: (3.2.1) priming of the microfluidic cassette, (3.2.2) blood separation, and (3.2.3) cell harvesting. To ensure optimal capture efficiency and reproducibility, follow these procedures in order.

3.2.1 Priming the Cassette

1. At the user display (label 1 in Fig. 2), select the priming protocol. This protocol prepares the instrument and a new cassette to receive a blood sample. On the Parsortix menu, choose protocol **PX2_PF**.
2. Initiate the run by pressing **[Run]** and then **[Start]**.
3. When prompted, load a new GEN3P6.5 6.5 μm separation cassette in the cassette clamp (label 2 in Fig. 2) and press **[OK]** to begin the priming. This process takes approximately 15 min to complete.
4. After the priming cycle ends, press **[OK]**, then **[Continue]** to return to the main menu (*see* **Note 9**).

3.2.2 Blood Separation

Before starting, make sure an empty 50 mL Falcon tube is attached to the instrument.

1. In the main menu, select the separation protocol PX2_S99F, press **[Run]**, and then **[Start]** to initiate separation. In this protocol, blood flows inside the cassette at a controlled and constant pressure of 99 mbar to enable cell separation.

1 User display
2 Cassette clamp
3 Waste bottle
4 Buffer bottle
5 Sample tube
6 Reagents tubes
7 Harvest/Separation valve
8 Harvest line & collection tube

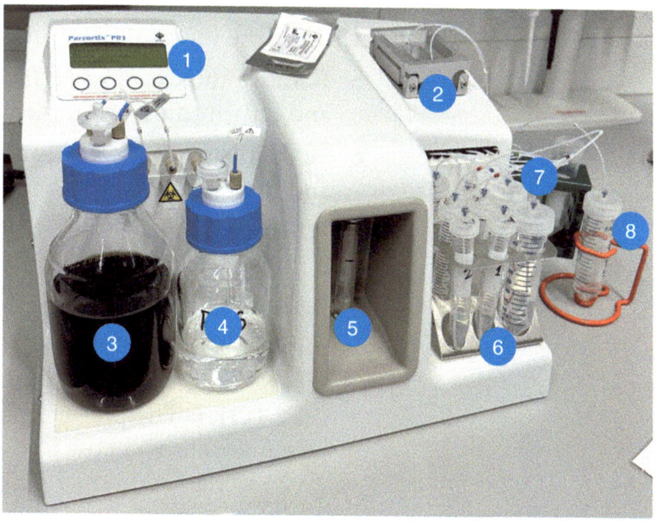

Fig. 2 Overview of the Parsortix PR1 instrument. Labeled components include: (1) user display for operation control, (2) cassette clamp for securing the separation cassette, (3) waste bottle, (4) buffer bottle, (5) sample tube compartment, (6) reagent tubes, (7) harvest/separation valve, and (8) harvest line connected to the collection tube. This configuration supports the automated processing of biological samples for downstream applications

2. At the "Rinse vacutainer" prompt, partially remove the attached tube (label 5 in Fig. 2) and press **[OK]**. Collect rinse fluid in the same tube. Avoid flicking the fluid from the line tip.
3. Gently invert the blood sample tube to resuspend the cells, mount the tube in the system, and press **[OK]** at the "Attach vacutainer" prompt (*see* **Notes 10** and **11**).
4. At "Start?" press **[OK]** to commence. This process takes approximately 60–90 min for a 3 mL sample.
5. Every ~20 min, gently tap the inclined vacutainer to resuspend settled cells and prevent clogging (*see* **Note 12**).
6. When the separation process is completed, press **[OK]** and then **[Continue]**.

3.2.3 Cell Harvesting

This step allows the removal of the cells captured inside the cassette and their collection for downstream genomic analysis.

1. Start by running the pre-harvesting protocol, selecting the protocol **PX2_CT2**, pressing **[Run],** and then **[Start]**.
2. At "Insert cleaning cassette," remove the separation cassette from the clamp, insert a GEN3P10 10 μm cleaning cassette, and press **[OK]** (*see* **Note 13**).
3. At "Empty rgt tubes," verify that all reagent tubes are empty (label 6 in Fig. 2). Press **[OK]** to start the flushing to remove residual blood cells before harvest. The process requires approximately 25 min to complete.
4. After the flush finishes, press **[OK]** and then **[Continue]**.
5. Remove the cleaning cassette, reinsert the separation cassette, and return to the main menu.
6. Select the harvest protocol by choosing **PX2_H**, press **[Run]**, and then **[Start]**.
7. When prompted, rotate the harvest valve anticlockwise (label 7 in Fig. 2) to "HAR" and press **[OK]**.
8. Remove the harvest line from the waste tube (label 8 in Fig. 2), wipe it with alcohol, and allow it to dry. Then, position a low-attachment 1.5 mL tube beneath it.
9. Press **[OK]**. The pre-set volume (e.g., 200 μL) containing the captured cells will flow into the collection tube (*see* **Note 14**). To collect an additional 1 mL, press **[YES]**.
10. Upon complete harvesting, rotate the harvest valve back to "SEP," press **[OK]**, return the harvest line to the waste tube, press **[OK]**, and then **[Continue]**.
11. It is recommended, at this point, to fix the cells to prevent potential cell or DNA damage (as blood samples were collected

in EDTA tubes). Add 200 µL of 4% PFA to the cell suspension (final PFA concentration is 2%), gently pipette to mix the solution, and incubate for 20 min at room temperature.

12. Upon incubation, fill the 1.5 mL tube with PBS and centrifuge at 400 g for 10 min at room temperature.
13. Carefully remove the supernatant without disturbing the cell pellet.
14. Wash the cell pellet with 1 mL of PBS and repeat centrifugation as in **step 12**.
15. Discard the supernatant and resuspend the cell pellet in 200 µL of PBS.
16. Optional step: CTC identification by immunostaining (*see* **Note 15**).

3.3 Single CTC Isolation by Micromanipulation

Before CTC isolation, we should perform the micromanipulator setup, which implies preparing the collection chamber and the capillary.

3.3.1 Preparation of the Collection Chamber for Cellular Suspension

1. Prepare a 5% BSA w/v solution by weighing 500 mg of BSA and dissolving it in 10 mL of PBS. Mix the solution thoroughly until the BSA is completely dissolved, then filter it using a 0.22 µm pore filter. Take a sterile, chambered cell culture slide and prepare it by pipetting 100 µL of the 5% BSA solution into each chamber that will be used. Then, remove it quickly (*see* **Note 16**).
2. Clearly distinguish at least two chambers on the slide, one designated as the picking field and the other(s) as the sample field (Fig. 3, step 1).
3. Pipette 200 µL of PBS into the picking field (*see* **Note 17**).
4. Pipette the 200 µL of the cell suspension into the sample field (Fig. 3, step 1). The maximum concentration of cells in the sample field should not exceed 1000 cells per chamber. These numbers will ensure an easy and efficient single CTC isolation without contaminating cells being carried over (Fig. 3) (*see* **Note 18**).
5. Place the slide in the fluorescence microscope specimen holder.
6. Allow the cells to settle at the bottom of the sample field for a few minutes.
7. Check under the microscope that all cells are sitting at the bottom of the slide (*see* **Note 19**).

3.3.2 Capillary Preparation for Micromanipulation

1. Pipette 100 µL of FBS into a sterile 1.5 mL reaction tube.
2. Submerge the capillary tip in the FBS, allowing it to be filled by capillary action.

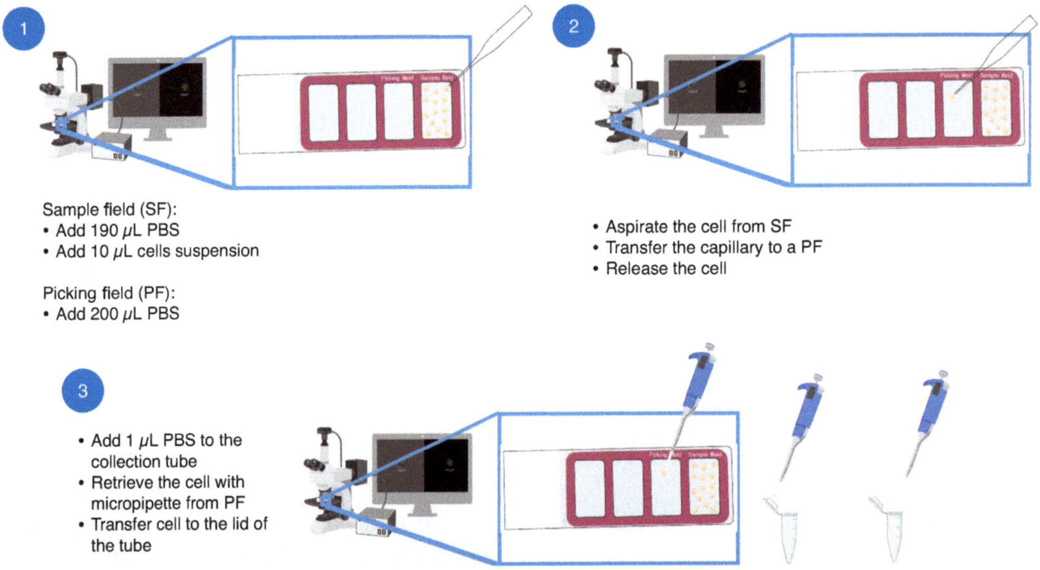

Fig. 3 Schematic illustration of the key steps involved in the single CTC micromanipulation process. A capillary visualized under a microscope is used to aspirate the target cell. The cell is transferred to a clean field, visually inspected for purity and integrity, and finally retrieved using a micropipette for transferring into a 0.2 mL PCR tube for downstream analysis

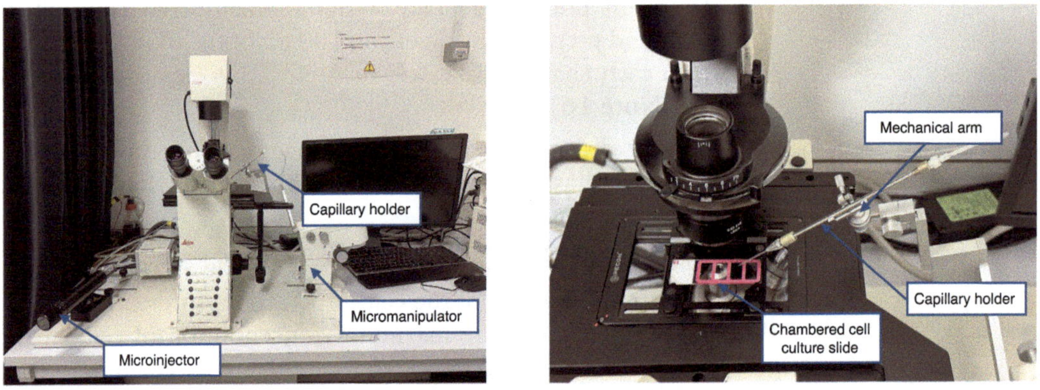

Fig. 4 Setup of the equipment used for single-cell micromanipulation, including a mechanical microinjector, a manual micromanipulator, and a fluorescence microscope, as illustrated in the images

3. Mount the capillary in the microinjector's capillary holder and attach the assembly to the micromanipulator's mechanical arm (Fig. 4—*see* **Note 20**).
4. Using the micromanipulator knobs, lower the capillary into the sample area until its tip just grazes the slide's surface (*see* **Note 21**). For best visibility and focus, use the 10X objective.
5. Once the capillary is positioned inside the sample field, eject a small amount of FBS from the capillary, then draw up 2–5 μL of PBS using the microinjector control knob. Always keep the capillary tip filled with liquid.

3.3.3 Cell Picking

Ensure the micromanipulation setup is fully prepared before beginning the cell-picking process (Fig. 4).

1. Place 1 μL of PBS on the lid of a sterile 0.2 mL PCR tube (*see* **Note 22**). This will be the collection tube.
2. Switch to a 20X objective to visualize the capillary and target cells clearly from the sample field.
3. Identify the cell of interest within the sample field (*see* **Note 23**), then position the capillary tip beside it using the control knobs, taking care not to push or dislodge the cell.
4. Clear nearby debris by aspirating and flushing PBS through the capillary, thereby washing away unwanted cells.
5. Carefully draw the target cell into the capillary; confirm its disappearance from view before retracting.
6. Withdraw the capillary, move it to the picking area, and expel the cell (Fig. 3, step 2; *see* **Note 24**).
7. Inspect under the microscope to verify that only the desired cell is present; if debris remains, repeat the PBS wash (as in step 4; *see* **Note 25**).
8. Fully retract the capillary from the picking field.
9. Photograph the isolated cell and its surroundings to document integrity and purity.
10. Position the tip of a 0.2–2 μL micropipette next to the cell, set it to 1 μL, and gently draw the cell into the pipette (Fig. 3, step 3; *see* **Note 26**).
11. Dispense the 1 μL containing the cell into the preloaded 0.2 mL PCR-tube lid by touching the pipette tip to the liquid drop (Fig. 3, step 3; *see* **Note 27**).
12. Place the lid on the fluorescence microscope to confirm the cell's presence (*see* **Note 28**).
13. Once the presence of the cell is confirmed, close the tube, briefly centrifuge to collect the liquid at the bottom, and store at −80 °C until amplification.

3.4 Single CTC WGA

Before Starting

1. All reagents should be thawed on ice, vortexed, and briefly spun down before use, except for Reagent 9, which should be thawed at room temperature. Enzymes should be removed from the freezer just before use, spun down, and added as the final components of the mix.
2. When handling high-density reagents, pipette slowly, mix by aspirating and dispensing several times, and ensure complete and accurate transfer of the entire volume into the mix.

3. All mixes must be freshly prepared except the preannealing reaction mix (*see* **Note 29**).
4. AMPure XP beads must be at room temperature for at least 30 min before use. Make sure they are homogeneously resuspended before pipetting.
5. 70% ethanol must be freshly prepared.
6. Always consider a 10% reagent extra volume for mix preparation (*see* **Note 30**).
7. Prepare single-cell amplifications in a laminar flow hood within a pre-PCR area, using a PCR-cooler, a dedicated set of pipettes, and UV-irradiated plastic materials.
8. All procedures from step 3.5 onward should be performed in a post-PCR area, using a dedicated set of pipettes, a PCR cooler, UV-irradiated plastic materials, and a PCR cabinet.
9. All sample tubes must be spun down before and after running the thermal cycler programs, and the thermal cycler lid should always be ON at 105 °C.
10. Keep sample tubes and mixes in a PCR cooler when processing.

Pre-PCR Area

3.4.1 Prepare CTCs and Controls for WGA

1. Remove single CTCs from −80 °C and thaw on ice for 10 min.
2. Spin down the tubes to recover the entire content at the bottom, and keep them in a PCR cooler while preparing the preannealing mix and the cell lysis reaction.
3. A positive and a negative WGA control tube should be included by adding 1 μL of 10 ng/μL gDNA (REPLI-g human control kit) and 1 μL of water (included in the kit), respectively (*see* **Note 31**).

3.4.2 Preannealing Mix

1. Mix 0.5 μL of each reagent (Reaction Buffer 1, Reagent 4, and Reagent 5) in a 1.5 mL LoBind tube and 1.5 μL of water per sample, taking into account a 10% extra volume (*see* **Note 30**) for the preannealing (PA) mix preparation. The reaction volume per sample is 3 μL.
2. Slightly vortex the mix and pulse it (*see* **Note 32**).
3. Run the preannealing program (Table 1).

3.4.3 Cell Lysis Reaction

1. Prepare the mix in a LoBind tube combining the following volumes of reagents per sample: 0.2 μL Reaction Buffer 1 + 0.13 μL Reagent 2 + 0.13 μL Reagent 3 + 0.26 μL Enzyme 1 + 1.28 μL of water. Take into consideration the 10% extra volume (*see* **Note 30**).
2. Vortex, spin down the mix, and add 2 uL of the mix per well (*see* **Note 33**).

Table 1
Preannealing program conditions. Number of cycles, temperature, and time conditions for running the preannealing program

# cycles	Temperature (°C)	Time (min)
1	65 ⇒ 15[a]	1
1	15	∞

[a]Start at 65 °C for 1 min and decrease the temperature 1 °C per minute until it reaches 15 °C. Total run time ~52 min

3. Spin down the sample tubes and run the cell lysis program: 56 °C for 15 min; 80 °C for 10 min; 4 °C ∞. The final volume per tube is 3 μL, and the total run time is ~26 min.

3.4.4 DNA Digestion

1. Prepare the DNA digestion mix in a 1.5 mL LoBind tube by combining the following reagents per sample: 0.2 μL Reaction Buffer 1 + 0.2 μL Enzyme 2 + 1.6 μL of water, considering a 10% extra volume (*see* **Note 30**).
2. Vortex, spin down the mix, and add 2 uL of the mix per well (*see* **Note 33**).
3. Spin down the sample tubes and run the DNA digestion program: 37 °C for 5 min; 65 °C for 5 min; 4 °C ∞. The final volume per tube is 5 μL, and the total run time is ~12 min.

3.4.5 Ligation

1. Prepare the ligation reaction mix directly into the PA product tube from Subheading 3.4.2 to avoid losing reagent volume when pipetting. Add 1 μL of Reagent 6 + 1 μL of Enzyme 3 per reaction.
2. Slightly vortex the tube and add 5 μL of the reaction mix to each tube containing the DNA digestion product (*see* **Note 33**).
3. Spin down the sample tubes and run the Ligation program: 15 °C for 40 min; 15 °C ∞. The final volume per tube is 10 μL, and the total run time is ~40 min.

3.4.6 Primary PCR

1. Mix the components in a 1.5 mL LoBind tube according to the indications in Table 2, taking into account a 10% extra volume (*see* **Note 30**).
2. Vortex and spin down the mix.
3. Add 40 μL per reaction (*see* **Note 33**) and briefly spin down the tubes before running the Primary PCR program: 68 °C for 3 min; 98 °C for 30 s; (94 °C for 15 s; 60 °C for 30 s; 68 °C for 4 min 36 s) 11 cycles; (94 °C for 15 s; 60 °C for 30 s; 68 °C for 6 min) 6 cycles; 68 °C for 5 min; 4 °C ∞. The final volume per sample is 50 μL, and the total run time is ~2 hours.

Table 2
Primary PCR reaction mix. Volume of each reagent needed to prepare the primary PCR reaction mix per sample

Reagent	Volume per reaction (μL)
Reaction 7	3
Reagent 8	2.5
Reagent 9	1
Enzyme 4	0.5
Water	33

Safe Stopping Point
The amplified products can be stored at −20 °C.

3.5 WGA QC

Post-PCR area

1. Prepare the QC mix by combining 6.5 μL PCR water + 1 μL PCR Reaction Buffer + 0.2 μL PCR dNTPs + 1 μL QC Primer Mix + 0.2 μL PCR BSA + 0.1 μL PCR Taq Polymerase per reaction, considering a 10% extra volume (*see* **Note 30**).
2. Vortex and spin the QC mix and dispense 9 μL per well.
3. Add 1 μL of the WGA product using a multichannel pipette, and include, again, a positive and a negative control sample using 1 μL of 100 ng/μL gDNA and 1 μL of PCR water, respectively.
4. Spin down all the tubes and run the WGA QC program: 95 °C for 4 min; (95 °C for 30 s; 58 °C for 30 s; 72 °C for 90 s) 32 cycles; 72 °C for 7 min; 4 °C ∞. The final volume per tube is 10 μL, and the total run time is ~2 h.

Safe Stopping Point
The WGA QC products can be stored at −20 °C.

3.6 QC Result Evaluation

To check the WGA QC products, use the Agilent TapeStation system with the D1000 Assay.

1. Allow reagents (sample buffer, ladder, and ScreenTape device) to equilibrate at room temperature for at least 30 min.
2. Vortex the D1000 sample buffer and spin it down together with the D1000 ladder tube.
3. Prepare the ladder well by adding 3 μL of sample buffer and 1 μL of the ladder to the first position (*see* **Note 34**).

4. Prepare sample wells by adding 3 μL of sample buffer and 1 μL of DNA sample (QC product).
5. Spin down and vortex for 1 min at 2000 rpm to facilitate mixing of the reagents.
6. Spin down the samples before loading them into the TapeStation system.
7. After running the TapeStation, check the gel and the electropherograms, and select those samples that are positive for at least three markers (Fig. 5).

3.7 Adaptor Removal

1. Prepare the reaction mix in a 1.5 mL LoBind tube by combining 5 μL NEBuffer 4 10X + 1 μL MseI 50 U/μL per reaction (*see* **Note 30**).
2. Vortex, spin it down, and dispense 6 μL of the mix per well.
3. Transfer 44 μL of WGA product from the selected samples based on QC results using a multichannel pipette. Then, run the adaptor removal program: 37 °C for 3 h; 65 °C for 20 min. The final volume per tube is 50 μL, and the total run time is ~3 h 22 min.

Safe Stopping Point

The adaptor removal products can be stored at −20 °C.

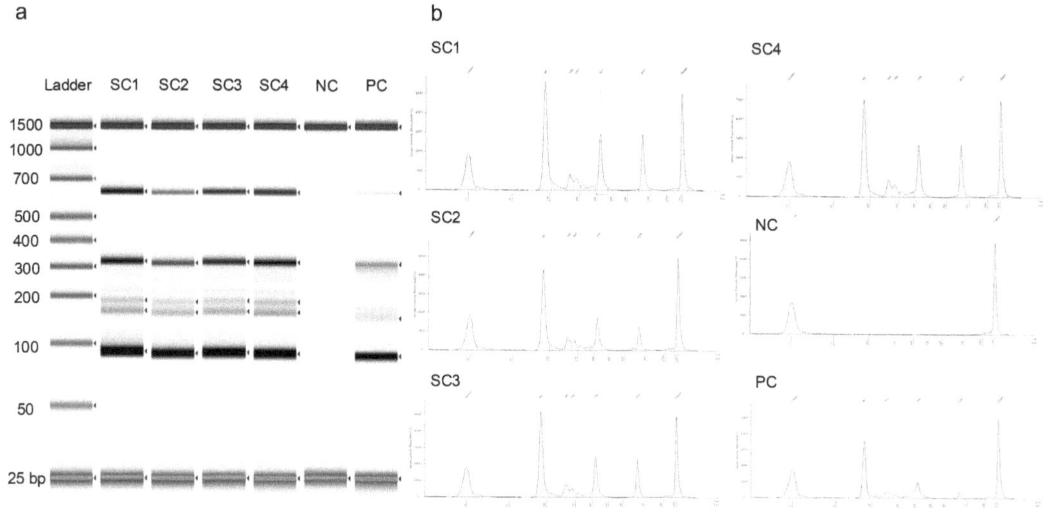

Fig. 5 Ampli1 Plus WGA QC products run on Agilent 2200 TapeStation system (Agilent Technologies). (**a**) Gel image of four single CTCs (SC1–SC4), negative control, and positive control. The four cells are positive for the four markers. (**b**) Electropherograms were obtained for the four single CTCs, as well as the negative and positive controls. Peaks correspond with the expected size. The marker, ~166 base pairs (bp), maps to a polymorphic region. Two bands mean heterozygosity (single CTCs), and one band means homozygosity (positive control). SC single cell, NC negative control, and PC positive control. Samples were loaded onto an Agilent D1000 ScreenTape

3.8 Sample Purification

1. Mix 90 μL of AMPure XP beads with the total volume (50 μL) of the adaptor removal product (beads ratio 1.8X) (*see* **Note 35**), and pipette up and down several times until the mixture appears homogeneous.
2. Incubate the mixed samples for 5 min at room temperature.
3. Place on the magnet for 10 min and remove the supernatant.
4. Keep the tubes on the magnet and wash the beads using 200 μL of 70% Ethanol (EtOH) per sample. Incubate for 30 s at room temperature and discard. Repeat this step for a total of two washes (*see* **Note 36**).
5. Let the sample tubes dry in the magnet for ~2 min (*see* **Note 37**).
6. Add 40 μL of NFW and mix by pipetting to resuspend the beads and elute DNA (*see* **Note 38**).
7. Incubate for 5 min at room temperature and place on the magnet until the supernatant becomes completely clear (3–5 min).
8. Transfer the eluted DNA (~38 μL) to a clean LoBind tube properly labeled (*see* **Note 39**).

3.9 Purified Ampli1 Plus Product Quantification Using the Qubit dsDNA BR Assay

1. Ensure all reagents are at room temperature for at least 30 min prior to use. Prepare the working solution by mixing 199 μL of Qubit buffer and 1 μL of Qubit reagent per sample, and then vortex the tube (*see* **Notes 40** and **41**).
2. Standards and samples must be prepared in the Qubit assay tubes. Prepare two standard tubes by adding 190 μL working solution + 10 μL standard 1 or 2, respectively (*see* **Note 42**).
3. Prepare sample tubes by adding 198 μL working solution + 2 μL cleaned-up DNA.
4. Slightly vortex the tubes for 2–3 s (*see* **Note 43**) and incubate at room temperature for 2 min.
5. Insert the tubes into the Qubit fluorometer according to the software instructions, and then, take the readings. Each scWGA reaction typically yields ~900–1000 ng of DNA.

3.10 Purified Ampli1 Plus Product Fragment Size Evaluation Using the TapeStation D1000 Assay

Repeat the procedure described in Subheading 3.6. The expected fragment size distribution ranges from 200 bp to >1500 bp (Fig. 6).

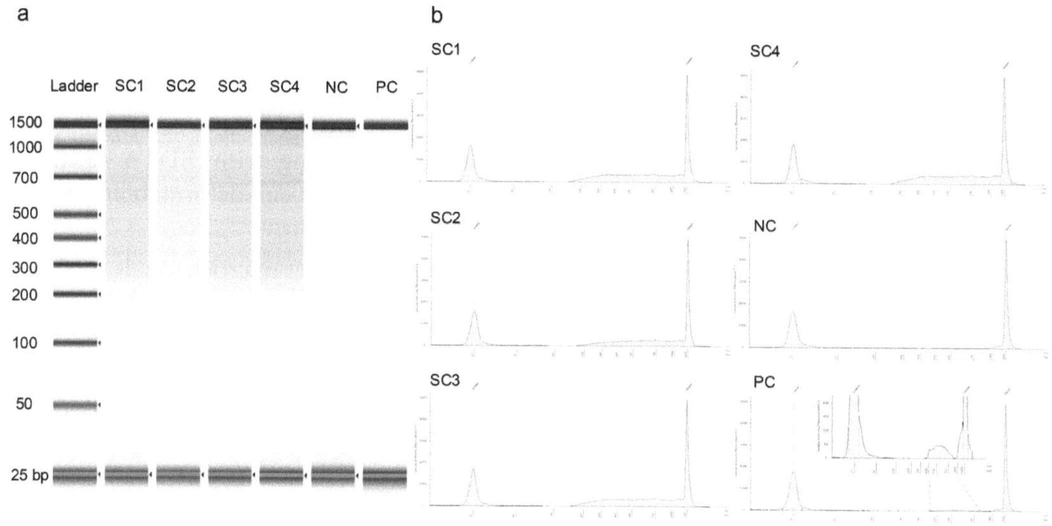

Fig. 6 Fragment size distribution after Ampli1 Plus product purification run on Agilent 2200 TapeStation system (Agilent Technologies). (**a**) Gel image for the four single CTCs (SC1-SC4), negative control, and positive control. (**b**) Electropherograms were obtained for the four single CTCs, negative, and positive controls. The fragment size distribution corresponds to the expected size: ~ 200–2000 bp. SC single cell, NC negative control, and PC positive control. Samples were loaded onto an Agilent D1000 ScreenTape

4 Notes

1. Coagulation risk and time sensitivity issues must be taken into consideration. Under anesthesia, larger volumes are obtained. Postmortem puncture not only risks rapid clotting but also yields less blood volume—perform collection within seconds to maximize volume and minimize coagulation.
2. Two major cardiac routes exist: Sub-xiphoid (diaphragmatic): the needle enters below the sternum through the diaphragm. Lateral intercostal: Enters at the left chest wall at elbow level. Chosen here for ease of access and consistency in anesthetized mice.
3. The mouse heart is positioned tilted to the left. Shaving just below the sternum on the left side exposes the beating apex.
4. Loss of pedal (toe-pinch) and tail-pinch reflexes confirm a surgical plane. If the mouse moves with needle placement, deepen anesthesia before proceeding.
5. Inserting the needle bevel up reduces tissue tearing, and inserting it perpendicularly at left elbow level targets the left ventricular apex. Avoid angling too shallowly or too steeply to prevent septal puncture.
6. A slight further advance from a bright-red flashback can sample right-ventricular blood but do so cautiously.

7. To prevent hemolysis, aspirate slowly to minimize shear forces and remove the needle before transferring to the tube to prevent cell lysis.
8. Blood preservatives, such as fixatives, can alter cell rigidity and reduce capture efficiency. Preloading with EDTA or using EDTA-coated tubes preserves native cell deformability, ensuring optimal CTC recovery and viability.
9. Cassette priming fully wets microchannels, clears manufacturing debris, and equilibrates flow dynamics—critical for reproducibility and maximal capture.
10. With a limited volume of mouse blood (<1 mL), dilute samples in PBS to a final volume of up to 3 mL to ensure adequate volume for processing. Rinse the collection and transfer tubes with PBS to recover any residual blood and prevent sample loss.
11. Parsortix is compatible with classical 10 mL EDTA Vacutainer tubes or 50 mL Falcon conical tubes; use tubes that fit securely on the instrument's sample ports.
12. Gently tapping the sample tube every 15–20 min prevents cell sedimentation and microchannel clogs. Avoid vigorous shaking.
13. The GEN3P10 10 μm cleaning cassette flushes residual cells and debris from the system before harvest, ensuring the purity of the collected fraction.
14. Our protocol uses 200 μL as the collection volume. Using small elution volumes helps concentrate rare cells and eliminates the need for centrifugation steps, thereby reducing the potential for cell loss and contamination. Reserve larger volumes only if a high yield is needed for downstream assays. To adjust the collection volume, speak to the Parsortix manufacturer.
15. In some cases, users may perform an immunostaining step—such as targeting cell surface epithelial markers—to identify CTCs and distinguish them from contaminating leukocytes. This protocol has been developed using a cell line that endogenously expresses eGFP, allowing for the easy identification of tumor cells.
16. Coating with BSA reduces undesirable cell adhesion to the glass surface.
17. Adjust the fluid volume in the chamber to facilitate manipulation, making sure there is sufficient media to work in.
18. The number of recoverable cells from blood samples of cancer patients varies depending on factors such as tumor type, tumor cell abundance, sample volume, and the efficiency of cell

enrichment. If the number of recovered cells is high, it is advisable to dilute the sample with PBS (e.g., pipette 190 μL of PBS into the sample field and 10 μL of cell suspension).

19. Check that your cell suspension density is suitable for single-cell isolation; dilute further if necessary.
20. This procedure utilizes a manual micromanipulator and mechanical microinjector (Fig. 4); however, automated platforms could also be used—follow the manufacturer's setup guide.
21. Glass capillaries bend rather than break when gently contacting the slide.
22. Rather than PBS, a cell lysis or specialized storage buffer can be loaded into the capillary.
23. Take photographs of each target cell before extraction to assess any morphological changes or damage that may have occurred during the process.
24. Dispense only a minimal amount of PBS into the picking region to avoid displacing the isolated cell.
25. If cells remain clumped after digestion, use the capillary tip as a micro-scalpel to carefully sever unwanted attachments.
26. Use a micropipette for final cell transfer to minimize contamination from residual debris in the capillary. Replacing capillaries for each cell increases handling time and may compromise cell quality.
27. Touching the pipette tip to the liquid ensures the cell is moved into the collection vessel.
28. Placing the cell in the tube lid rather than its bottom facilitates its visibility under the microscope.
29. If there is only one thermal cycler available, run the preannealing (PA) program in advance and store the product at −20 °C for up to 1 week. Thaw it on ice, vortex, and spin before use.
30. To compensate for pipetting loss when preparing reaction mixes, add a 10% extra volume. This means that for 20 reactions (18 CTCs + 2 controls), the volume needed for each reagent should be as for 22 reactions.
31. To prevent carryover contamination, the negative control should be processed first, followed by the samples, and the positive control should be processed last.
32. If the PA mix volume exceeds 100 μL, split the mix into two tubes of equal volume using 0.2 mL PCR tubes.
33. Always pipette the mixes onto the tube wall. Do not touch the liquid inside the tube, as it can cause the loss of material.

34. The ladder should be run as the first well for the fragments reference. When pipetting the sample buffer, ensure that excess droplets are removed from the tip before transferring it to the sample well.
35. Beads can be added directly to the sample tubes after the adaptor removal step. When handling a large number of samples, a reservoir and a multichannel pipette can be used to speed up the process. Mixing by pipetting provides more reproducible results than vortexing.
36. To properly clean the beads, let the 70% EtOH run down the tube wall where the beads are stuck.
37. Keep the sample tubes open on the magnet and avoid overdrying the beads. Don't let them crack, as this can lead to sample loss.
38. It is recommended to add the NFW to all tubes simultaneously using a multichannel pipette. This may provide more reproducibility, avoiding differences in drying times among samples, especially when working with a large batch.
39. Leaving a few μL of NFW in the sample tube doesn't compromise the yield and helps prevent bead carryover, which can interfere with sequencing protocols.
40. Protect the Qubit working solution tube from light with aluminum foil to avoid reagent degradation (it is a fluorochrome).
41. Vortex the working solution by gradually increasing the speed, then decreasing it back to zero, to ensure thorough mixing while minimizing foam and bubble formation.
42. Label the tubes on the cap only, as any markings on the tube wall may interfere with concentration measurements.
43. It is highly recommended to protect the bottom of the tubes with a tissue or glove when vortexing to avoid contact with the rubber pad, which can leave residues on the tube edge and potentially interfere with concentration measurements.

Acknowledgments

This work was supported by the Spanish MICINN grant PID2019-106247GB-100 awarded to D.P.

References

1. Navin NE (2014) Cancer genomics: one cell at a time. Genome Biol 15(8):452
2. Zhang L, Cui X, Schmitt K, Hubert R, Navidi W, Arnheim N (1992) Whole genome amplification from a single cell: implications for genetic analysis. Proc Natl Acad Sci USA 89(13):5847–5851
3. Klein CA, Schmidt-Kittler O, Schardt JA, Pantel K, Speicher MR, Riethmüller G (1999) Comparative genomic hybridization, loss of

heterozygosity, and DNA sequence analysis of single cells. Proc Natl Acad Sci USA 96(8): 4494–4499

4. Dean FB, Hosono S, Fang L, Wu X, Faruqi AF, Bray-Ward P et al (2002) Comprehensive human genome amplification using multiple displacement amplification. Proc Natl Acad Sci USA 99(8):5261–5266
5. Picher ÁJ, Budeus B, Wafzig O, Krüger C, García-Gómez S, Martínez-Jiménez MI et al (2016) TruePrime is a novel method for whole-genome amplification from single cells based on TthPrimPol. Nat Commun 7:13296
6. Gonzalez-Pena V, Natarajan S, Xia Y, Klein D, Carter R, Pang Y et al (2021) Accurate genomic variant detection in single cells with primary template-directed amplification. Proc Natl Acad Sci USA 118(24)
7. Kamberov E, Sun T, Bruening E, Pinter JH, Sleptsova I, Kurihara T et al (2012) Amplification and analysis of whole genome and whole transcriptome libraries generated by a DNA polymerization process. Patent
8. Zong C, Lu S, Chapman AR, Xie XS (2012) Genome-wide detection of single-nucleotide and copy-number variations of a single human cell. Science 338(6114):1622–1626
9. Estévez-Gómez N, Prieto T, Tomás L, Alvariño P, Guillaumet-Adkins A, Heyn H et al (2025) Differential performance of strategies for single-cell whole-genome amplification. Cell Rep Methods 5(4):101025
10. Khoo BL, Chaudhuri PK, Lim CT, Warkiani ME (2017) Advancing techniques and insights in circulating tumor cell (CTC) research. In: Cancer drug discovery and development. Springer, Cham, pp 71–94
11. Agashe R, Kurzrock R (2020) Circulating tumor cells: from the laboratory to the cancer clinic. Cancers (Basel) 12(9):2361
12. Chéry L, Lam HM, Coleman I, Lakely B, Coleman R, Larson S et al (2014) Characterization of single disseminated prostate cancer cells reveals tumor cell heterogeneity and identifies dormancy associated pathways. Oncotarget 5(20):9939–9951
13. Krebs MG, Metcalf RL, Carter L, Brady G, Blackhall FH, Dive C (2014) Molecular analysis of circulating tumour cells-biology and biomarkers. Nat Rev Clin Oncol 11(3):129–144
14. Alves JM, Estévez-Gómez N, Valecha M, Prado-López S, Tomás L, Alvariño P et al (2022) Comparative analysis of capture methods for genomic profiling of circulating tumor cells in colorectal cancer. Genomics 114(6): 110500
15. Babayan A, Alawi M, Gormley M, Müller V, Wikman H, McMullin RP et al (2017) Comparative study of whole genome amplification and next generation sequencing performance of single cancer cells. Oncotarget 8(34): 56066–56080
16. Carpenter EL, Rader J, Ruden J, Rappaport EF, Hunter KN, Hallberg PL et al (2014) Dielectrophoretic capture and genetic analysis of single neuroblastoma tumor cells. Front Oncol 4(July):1–14

Chapter 6

An Integrated Preimplantation Genetic Testing Approach with MALBAC-Based Whole-Genome Sequencing

Cheng Wan, Fangfang Gao, and Sijia Lu

Abstract

Preimplantation genetic testing (PGT) is crucial for selecting embryos free of genetic abnormalities. However, existing PGT approaches often necessitate separate platforms for aneuploidy (PGT-A), monogenic disorders (PGT-M), and structural rearrangements (PGT-SR). This can drive up costs and operational complexity when multiple PGT tests are required for a single embryo. Here, we present a MALBAC-based method that integrates PGT-A, PGT-M, and PGT-SR into one unified platform.

Key words Preimplantation genetic testing, MALBAC, Whole genome sequencing

1 Introduction

Preimplantation genetic testing (PGT) has revolutionized assisted reproductive technology since its inception in the early 1990s, offering a critical means of selecting embryos free of genetic abnormalities. Traditionally, PGT is categorized into three subtypes: PGT for aneuploidy (PGT-A), PGT for monogenic disorders (PGT-M), and PGT for structural rearrangements (PGT-SR) [1]. Each of these subtypes has been addressed through separate technical platforms, which has led to increased costs and operational complexity, especially when multiple tests are required for a single embryo [2, 3]. Studies have shown that a significant proportion of embryos deemed normal by PGT-M are later found to be aneuploid, underscoring the need for simultaneous PGT-M and PGT-A [4]. Similarly, patients undergoing PGT-SR often require the analysis of copy number variations (CNVs) and haplotypes, necessitating the integration of multiple analyses [5].

Over the past decades, several comprehensive PGT approaches have emerged, primarily based on single-nucleotide polymorphism (SNP) arrays and next-generation sequencing (NGS), including Karyomapping, OnePGT, Haploseek, low-depth whole-genome

Yafeng Ma (ed.), *Whole Genome Amplification: Methods and Protocols*, Methods in Molecular Biology, vol. 3003,
https://doi.org/10.1007/978-1-0716-5060-8_6,

sequencing (WGS)-based comprehensive PGT, GENType, and HaploPGT [6–10]. While these methods have shown promise, they often face limitations such as high costs, reduced sensitivity to mosaicism, and the need for higher sequencing depths.

Inspired by the relatively stable amplification bias of MALBAC [11], we developed KaryoSeq, a WGS-based comprehensive PGT approach for diverse genetic conditions. Building on the Mutated Allele Revealed by Sequencing with Aneuploidy and Linkage Analyses (Marasla) strategy [12], KaryoSeq enables simultaneous PGT-A, PGT-M, and PGT-SR, which requires genome-wide sequencing data at a depth of 2× (about 20 million raw reads) for most genes [13].

2 Materials

2.1 MALBAC Single Cell Whole GenomeAmplification

1. Sample preservation buffer: 30 mM Tris-HCl, pH 7.8, 2 mM EDTA, 20 mM KCl, 0.2% Triton X-100. Add about 100 mL double distilled water (ddH_2O) to a 1 L graduated cylinder. Weigh 3.63 g Tris, 0.58 g EDTA, 1.49 g KCl, and transfer to the cylinder. Add 2 mL Triton X-100 and about 800 mL ddH_2O. Mix and adjust pH with HCl. Make up to 1 L with ddH_2O. Store at −20 °C.
2. Lyase: 125 μg/mL protease. Weigh 2.5 mg protease to a 1.5 mL micro-centrifuge tube, add 1 mL ddH_2O to the tube. Store at −20 °C. Dilute 20 times with ddH_2O before use.
3. WGA buffer: 20 mM Tris-HCl, pH 8.8, 10 mM KCl, 10 mM NH_4SO_4, 3.5 mM $MgSO_4$, 0.1% Triton X-100, 0.4 mM each dNTP. Add 25 mL ddH_2O to a 100 mL graduated cylinder. Weigh 242.2 mg Tris, 74.6 mg KCl, 132.1 mg NH_4SO_4, and transfer to the cylinder. Add 0.1 mL Triton X-100 and about 50 mL ddH_2O. Mix and adjust pH with HCl. Add 4 mL 10 mM each dNTP solution, 0.32 mL Primer 1, 0.25 mL Primer 2, and 0.66 mL Primer 3. Make up to 100 mL with ddH_2O. Store at −20 °C.
4. Primer sequence (5′ to 3′):

   Primer 1: GTGAGTGATGGTTGAGGTAGTGTGGAGNNNNNGGG

   Primer 2: GTGAGTGATGGTTGAGGTAGTGTGGAGNNNNNTTT

   Primer 3: Biotin- GTGAGTGATGGTTGAGGTAGTGTGGAG

   Dissolve the primer powder to 100 μM with ddH_2O according to the provider's instructions. Store at −20 °C.
5. WGA enzyme: Deep Vent (exo-) DNA polymerase, 2 U/μL, New England Biolabs, catalogue number M0259.

2.2 Next-Generation Sequencing Library Preparation

1. NGS library preparation kit: The NGS library can be prepared using any verified commercial kit. Here, we use the NEBNext kit as an example.
   (a) NEBNext UltraShear DNA Fragmentation Module: New England Biolabs, M7634.
   (b) NEBNext Ultra II End Repair/dA-Tailing Module: New England Biolabs, catalogue number E7546.
   (c) NEBNext Ultra II Ligation Module: New England Biolabs, catalogue number E7595.
   (d) NEBNext Ultra II Q5 Master Mix: New England Biolabs, catalogue number M0544.
2. Ethanol: Add 80 mL absolute ethanol to a 100 mL graduated cylinder, make up to 100 mL with ddH2O. Store at 4 °C for no more than 3 days.
3. DNA purification magnetic beads: DNA purification and size selection can be performed using any verified commercial kit. Here, we use the AMPure XP beads from Beckman Coulter Life Science as an example.

3 Methods

3.1 WGA with MALBAC

1. Sample preparation.
   (a) Embryo cell samples: Collect 3–5 blastocyst trophectoderm cells in the embryo operation room, and place immediately at the bottom of a 0.2 mL PCR tube containing 5 μL sample preservation buffer. Store the sample(s) at −20 °C or −80 °C before use. This step should be performed by a well-trained embryologist.
   (b) Parental samples: Collect >5 mL peripheral blood with EDTA anticoagulation tube, extract genomic DNA from peripheral blood by using any verified commercial kit. Quantify the gDNA using the Qubit dsDNA HS Assay Kit. Store the samples at −20 °C before use.
   (c) Positive and blank controls: Dilute gDNA samples with known karyotypes to 50–100 pg/μL and use 1 μL as the positive control. Use DNA-free ultrapure water as the blank control.
2. Dilute the gDNA to 10 ng/μL, and add 1 μL to a 0.2 mL PCR tube with 4.5 μL sample preservation buffer. Thaw the embryo cell samples at room temperature.
3. Add 0.5 μL lyase onto the inner wall of samples, slightly above the liquid surface (do not insert the pipette tip into the liquid to prevent taking cells away). After that, centrifuge the tubes (30–60 s as recommended). Avoid air bubbles during operation and do not vortex or invert the tubes.

Table 1
Program for sample lysis

Cycle	Temperature	Time
1	50 °C 80 °C 4 °C	20 min 10 min Hold

Table 2
PCR program for whole genome amplification

Cycle	Temperature	Time
1	94 °C	3 min
8	10 °C 30 °C 50 °C 70 °C 95 °C	20 s 30 s 40 s 2 min 20 s
1	94 °C	30 s
17	94 °C 58 °C 72 °C	20 s 15 s 2 min
1	72 °C	5 min
1	4 °C	Hold

4. Perform thermal cycling of the samples with the heated lid turned ON, using the program in Table 1.
5. Stop the program when the lysis process is done. Proceed to the next step immediately.
6. Add 60 μL WGA buffer and 1 μL WGA enzyme to each sample, vortex the mixture thoroughly, and centrifuge briefly (micro-centrifuge, 2000 rpm, 3–5 s).
7. Perform thermal cycling of the samples with the heated lid turned ON, using the program in Table 2.
8. Stop the program when the process is done. Proceed to the next step immediately (*see* **Note 1**).
9. Purify the WGA products using the AMPure XP magnetic beads. Transfer 50 μL of WGA product into a new 1.5 mL micro-centrifuge tube, and add an equivalent volume of re-suspended AMPure XP beads into the same tube. Vortex and mix thoroughly, and then centrifuge briefly (microcentrifuge, 2000 rpm, 3–5 s). Incubate at room temperature for 5 min.

10. After incubation, place the tube on a magnetic rack for about 5 min. After the supernatant becomes clear, while keeping the tube on the magnetic rack, carefully aspirate and discard the supernatant. Do not touch the magnetic beads.
11. Add 200 μL freshly prepared 80% ethanol to the tube. Incubate at room temperature for 30 s, then carefully aspirate and discard the supernatant. DO NOT take the tube off the magnetic rack. DO NOT touch or disrupt the beads during this procedure.
12. Repeat **step 11** once.
13. Keep the tube open on the magnetic rack at room temperature for 3–5 min to evaporate the ethanol and make sure the beads are dried but not overdried.
14. Add 20 μL nuclease-free water and close the lids, then resuspend the beads by vortex (15–30 s). Pulse centrifuge the tubes, and incubate at room temperature for 5 min off the rack.
15. Place the tube onto the magnetic rack and wait for about 5 min till the supernatant becomes clear. Carefully transfer 18 μL supernatant to a new 1.5 mL centrifuge tube. DO NOT touch the beads during this procedure (*see* **Note 2**).
16. Quantify the gDNA using the Qubit dsDNA HS Assay Kit (*see* **Note 3**).

3.2 Enzymatic Fragmentation

1. Prepare the DNA fragmentation mix according to Table 3.
2. Add 1 μL of NEBNext UltraShear enzyme to the above mix, gently pipette for 6–8 times to mix evenly (do not vortex), centrifuge briefly, and place on ice for later use (*see* **Note 4**).
3. Perform thermal cycling of the samples with the temperature of the heated lid to 70 °C, tighten the heated lid, using the program outlined in Table 4 (*see* **Note 5**).
4. Stop the program when the process is done. Proceed to the next step immediately.

Table 3
Preparation of DNA fragmentation mix

Component	Volume
UltraShear reaction buffer	2.5 μL
Purified WGA product (50 ng)	X μL
Nuclease-free water	(6.5–X) μL
Total volume	*9 μL*

Table 4
Program for DNA fragmentation

Cycle	Temperature	Time
1	4 °C	1 min
1	32 °C	10 min
1	65 °C	15 min
1	4 °C	Hold

Table 5
Program for end repair and dA-Tailing

Cycle	Temperature	Time
1	20 °C	30 min
1	65 °C	30 min
1	4 °C	Hold

Table 6
Program for adapter ligation

Cycle	Temperature	Time
1	20 °C	15 min
1	4 °C	Hold

3.3 NGS Library Preparation

1. End repair and dA-Tailing. Add 3.5 μL End Prep Reaction Buffer, 1.5 μL End Prep Enzyme Mix, and 15 μL nuclease-free water to the fragmented DNA, the total volume is 30 μL. Vortex the mixture thoroughly and centrifuge briefly (micro-centrifuge, 2000 rpm, 3–5 s).
2. Perform thermal cycling of the samples with the temperature of the heated lid to 75 °C, and tighten the heated lid, using the program outlined in Table 5.
3. Adapter ligation. Add 15 μL Ligation Master Mix, 0.5 μL Ligation Enhancer, and 1.5 μL NGS library adapter (1.5 μM) to the End Prep Reaction Mixture from above step.
4. Perform thermal cycling of the samples with the heated lid turned OFF, using the program outlined in Table 6.
5. DNA size selection. Transfer the adapter ligated product into a new 1.5 mL micro-centrifuge tube, and add 23.5 μL (0.5× volume) of resuspended AMPure XP beads into the same

tube. Vortex and mix thoroughly, and then centrifuge briefly (microcentrifuge, 2000 rpm, 3–5 s). Incubate at room temperature for 5 min.

6. Place the tube on a magnetic stand for about 5 min. After the supernatant becomes clear, keeping the tube on the magnetic stand, carefully transfer the supernatant to a new 1.5 mL tube and discard the magnetic beads.
7. Add 7 μL (approximately 0.15× volume) of resuspended AMPure XP beads into the supernatant. Vortex and mix evenly, then centrifuge briefly (microcentrifuge, 2000 rpm, 3–5 s). Incubate at room temperature for 5 min.
8. Place the centrifuge tubes on a magnetic stand for about 5 min. When the supernatant becomes clear, while keeping the tube on the magnetic stand, carefully aspirate and discard the supernatant. Do not touch the magnetic beads.
9. Add 200 μL freshly prepared 80% ethanol to the tube. Incubate at room temperature for 30 s, then carefully aspirate and discard the supernatant. DO NOT take the tube off the magnetic rack. DO NOT touch or disrupt the beads during this procedure.
10. Repeat **step 9** once.
11. Keep the tube open on the magnetic rack at room temperature for 3–5 min to evaporate the ethanol, and make sure the beads are dried but not overdried.
12. Add 14 μL nuclease-free water and close the lids, then resuspend the beads by vortex (15–30 s). Pulse centrifuges the tubes, and incubate at room temperature for 5 min off the rack.
13. Place the tube onto the magnetic stand and wait for about 5 min till the supernatant becomes clear. Carefully transfer 11.5 μL supernatant to a new 0.2 mL PCR tube. DO NOT touch the beads during this procedure (*see* **Note 6**).
14. Library amplification. Add 12.5 μL Q5 Master Mix and 1 μL index primer to the size-selected DNA from above step (Fig. 1). Vortex thoroughly and centrifuge briefly (microcentrifuge, 2000 rpm, 3–5 s) (*see* **Note 7**).
15. Perform thermal cycling of the samples with the heated lid turned ON, using the program outlined in Table 7:
16. Library purification. Transfer 25 μL of PCR product into a new 1.5 mL micro centrifuge tube, and add an equivalent volume of re-suspended AMPure XP beads into the same tube. Vortex and mix thoroughly, and then centrifuge briefly (microcentrifuge, 2000 rpm, 3–5 s). Incubate at room temperature for 5 min.

NEBNext i501 Primer

5′- AATGATACGGCGACCACCGAGATCTACACTATAGCCTACACTCTTTCCCTACACGACGCTCTTCCGATC*T-3′

NEBNext i701 Primer

5′- CAAGCAGAAGACGGCATACGAGATCGAGTAATGTGACTGGAGTTCAGACGTGTGCTCTTCCGATC*T-3′

Fig. 1 A typical example of index primers from NEBNext kit. The index sequences are indicated with green letters, while the universal sequences are indicated with black letters

Table 7
PCR program for library amplification

Cycle	Temperature	Time
1	95 °C	30 s
10	95 °C 65 °C 72 °C	10 s 30 s 30 min
1	72 °C	5 min
1	4 °C	Hold

17. After incubation, place the tube on a magnetic rack for about 5 min. After the supernatant becomes clear, while keeping the tube on the magnetic rack, carefully aspirate and discard the supernatant. Do not touch the magnetic beads.
18. Add 200 μL freshly prepared 80% ethanol to the tube. Incubate at room temperature for 30 s, then carefully aspirate and discard the supernatant. DO NOT take the tube off the magnetic rack. DO NOT touch or disrupt the beads during this procedure.
19. Repeat **step 18** once.
20. Keep the tube open on the magnetic rack at room temperature for 3–5 min to evaporate the ethanol, and make sure the beads are dried but not overdried.
21. Add 20 μL nuclease-free water and close the lids, then resuspend the beads by vortex (15–30 s). Pulse centrifuges the tubes and incubate at room temperature for 5 min off the rack.
22. Place the tube onto the magnetic rack and wait for about 5 min till the supernatant becomes clear. Carefully transfer 18 μL supernatant to a new 1.5 mL centrifuge tube. DO NOT touch the beads during this procedure (*see* **Note 8**).
23. Quantify the library using the Qubit dsDNA HS Assay Kit.

3.4 NGS Sequencing

Pool the libraries and perform PE150 sequencing with high-throughput NGS platform according to the manufacturer's instructions. 20–30 million paired-end reads are required for each library.

4 Notes

1. We recommend that new users assess the quality of WGA products and amplified libraries by loading 5 μL of each sample onto a 1% agarose gel or using a bioanalyzer, following the manufacturer's instructions, to determine the library size. The fragment size of WGA products before fragmentation is expected to range from approximately 200 bp to 2000 bp (Fig. 2), while the size profile of the libraries should range from 300 bp to 600 bp (Fig. 3).
2. The purified WGA product can be stored at −20 °C for up to 1 month. For long-term storage, −80 °C storage is recommended. Try to avoid unnecessary freezing and thawing.
3. The concentration of purified WGA products should reach 10 ng/μL for the positive control, gDNA, and cell samples, while the concentration of the blank control WGA products should be less than 1 ng/μL.
4. Do not vortex during this step. Mix the reagents thoroughly by gently pipetting 6–8 times, centrifuge briefly, and place the mixture on the ice.

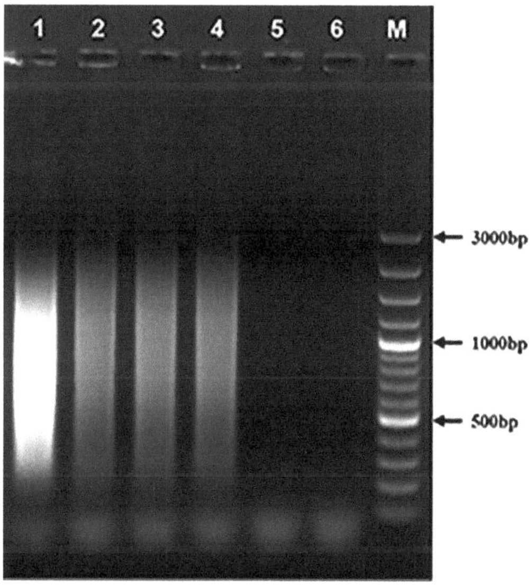

Fig. 2 The DNA size profile for MALBAC WGA products using 1% agarose gel electrophoresis. Lane 1: positive control; Lane 2–4: cell samples; Lane 5 & 6: blank controls; M: 100 bp plus DNA ladder

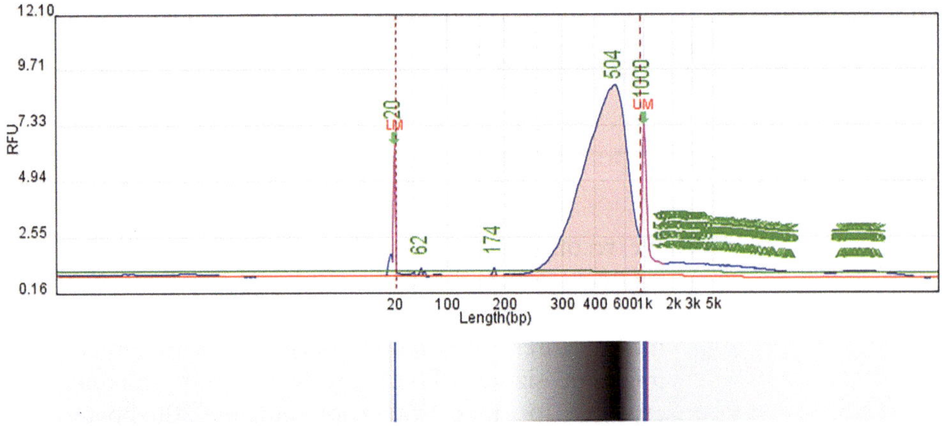

Fig. 3 The DNA size profile for a representative library using the Qsep400

5. Fragment time needs to be optimized according to the quantity of dsDNA. If there is a small deviation from the desired size, the reaction time can be adjusted as appropriate. It is recommended to set up a control with an extension of 3 min and a reduction of 3 min, which will help to determine the exact reaction time to the desired size.
6. The purified ligation product can be stored overnight at 4 °C or −20 °C. Long-term storage before purification is not recommended.
7. Assign unique barcodes to samples that will be sequenced in the same run or lane. Document the barcodes used with clear annotations. Adopt a meticulous approach to prevent cross-contamination when handling index tubes. If you come into contact with the inside of a tube or any of the index solution, change your gloves immediately to maintain sample integrity.
8. The purified library can be stored at −20 °C for up to 1 month. For long-term storage, −80 °C storage is recommended. Try to avoid unnecessary freezing and thawing, as sequencing data quality might be affected when they are frozen and thawed more than three times.

References

1. Zegers-Hochschild F, Adamson GD, Dyer S, Racowsky C, de Mouzon J, Sokol R et al (2017) The international glossary on infertility and fertility care, 2017. Hum Reprod 32(9): 1786–1801. https://doi.org/10.1093/humrep/dex234
2. Toft CLF, Ingerslev HJ, Kesmodel US, Diemer T, Degn B, Ernst A et al (2020) A systematic review on concurrent aneuploidy screening and preimplantation genetic testing for hereditary disorders: what is the prevalence of aneuploidy and is there a clinical effect from aneuploidy screening? Acta Obstet Gynecol Scand 99(6):696–706. https://doi.org/10.1111/aogs.13823
3. Zhang S, Lei C, Wu J, Xiao M, Zhou J, Zhu S et al (2021) A comprehensive and universal approach for embryo testing in patients with different genetic disorders. Clin Transl Med

11(7):e490. Epub 2021/07/30. https://doi.org/10.1002/ctm2.490

4. Minasi MG, Fiorentino F, Ruberti A, Biricik A, Cursio E, Cotroneo E et al (2017) Genetic diseases and aneuploidies can be detected with a single blastocyst biopsy: a successful clinical approach. Hum Reprod (Oxford, England) 32(8):1770–1777. Epub 2017/06/22. https://doi.org/10.1093/humrep/dex215
5. Hu L, Cheng D, Gong F, Lu C, Tan Y, Luo K et al (2016) Reciprocal translocation carrier diagnosis in preimplantation human embryos. EBioMedicine 14:139–147. Epub 2016/11/15. https://doi.org/10.1016/j.ebiom.2016.11.007
6. Xie P, Hu X, Kong L, Mao Y, Cheng D, Kang K et al (2022) A novel multifunctional haplotyping-based preimplantation genetic testing for different genetic conditions. Hum Reprod (Oxford, England). 37(11):2546–2559. Epub 2022/09/07. https://doi.org/10.1093/humrep/deac190
7. Masset H, Zamani Esteki M, Dimitriadou E, Dreesen J, Debrock S, Derhaag J et al (2019) Multi-centre evaluation of a comprehensive preimplantation genetic test through haplotyping-by-sequencing. Hum Reprod (Oxford, England). 34(8):1608–1619. Epub 2019/07/28. https://doi.org/10.1093/humrep/dez106
8. Backenroth D, Zahdeh F, Kling Y, Peretz A, Rosen T, Kort D et al (2019) Haploseek: a 24-hour all-in-one method for preimplantation genetic diagnosis (PGD) of monogenic disease and aneuploidy. Genetics Med 21(6):1390–1399. Epub 2018/11/20. https://doi.org/10.1038/s41436-018-0351-7
9. De Witte L, Raman L, Baetens M, De Koker A, Callewaert N, Symoens S et al (2022) GENType: all-in-one preimplantation genetic testing by pedigree haplotyping and copy number profiling suitable for third-party reproduction. Hum Reprod (Oxford, England) 37(7):1678–1691. Epub 2022/05/14. https://doi.org/10.1093/humrep/deac088
10. Chen S, Yin X, Zhang S, Xia J, Liu P, Xie P et al (2021) Comprehensive preimplantation genetic testing by massively parallel sequencing. Hum Reprod (Oxford, England). 36(1):236–247. Epub 2020/12/12. https://doi.org/10.1093/humrep/deaa269
11. Zong C, Lu S, Chapman AR, Xie XS (2012) Genome-wide detection of single-nucleotide and copy-number variations of a single human cell. Science (New York, NY). 338(6114):1622–1626. Epub 2012/12/22. https://doi.org/10.1126/science.1229164
12. Wu H, Shen X, Huang L, Zeng Y, Gao Y, Shao L et al (2018) Genotyping single-sperm cells by universal MARSALA enables the acquisition of linkage information for combined pre-implantation genetic diagnosis and genome screening. J Assist Reprod Genet 35(6):1071–1078. Epub 2018/05/24. https://doi.org/10.1007/s10815-018-1158-9
13. Li S, Chang C, Bai H, Qian W, Zou Y, Wu D et al (2025. Epub 2025/03/03) A novel and comprehensive whole-genome sequencing-based preimplantation genetic testing approach for different genetic conditions. J Mol Diagn. https://doi.org/10.1016/j.jmoldx.2025.02.002

Chapter 7

Handmade Biopsies and Whole Genome Amplification for Genotyping and Genomic Selection of Bovine Preimplantation Embryos

Clara Slade Oliveira, Luiz Sergio de Almeida Camargo, and Naiara Zoccal Saraiva

Abstract

The protocol describes the removal of embryonic cells using a simple manual method that eliminates the need for micromanipulation equipment, followed by sample processing for whole genome amplification, up to the point at which the samples are sent to the genotyping laboratory. Thus, steps required for genomic selection of bovine embryos—from the in vitro fertilization (IVF) laboratory to the genotyping service—are encompassed.

Grade I embryos at the blastocyst stage, preferably on day 7 (154–168 h.p.i.), are selected and transferred to biopsy plates containing identified microdroplets, with one embryo per droplet. Using an appropriate microblade, the operator cuts a portion of the trophectoderm from the embryo. The embryo is then cryopreserved, and the excised sample is transferred to a 0.2 mL tube and stored. The second stage of the protocol involves processing the sample by lysing cell membranes to release and denature DNA, followed by amplification of the DNA through an enzymatic reaction. These operations are carried out using a thermal cycler and, by the end of the protocol, yield a sufficient amount of DNA for submission to the genotyping laboratory.

Key words Embryo biopsy, Genotyping, Genomic selection, Bovine, IVF

1 Introduction

Genomic selection has become a cornerstone of genetic improvement in livestock. By identifying polymorphisms associated with performance traits, genotyping enables early and more accurate selection, enhancing overall system productivity. In recent decades, the adoption of genotyping in cattle breeding has accelerated, driven by declining costs and increasingly precise predictive models [1].

Yafeng Ma (ed.), *Whole Genome Amplification: Methods and Protocols*, Methods in Molecular Biology, vol. 3003, https://doi.org/10.1007/978-1-0716-5060-8_7,

Advances in dairy cattle breeding have extended genomic selection to embryo samples [1], allowing the genetic merit of individuals to be assessed prior to birth. This approach is particularly valuable in scenarios where herd reduction is necessary, such as in response to climate change and the imperative to lower methane emissions. By improving productivity per animal, it becomes possible to maintain output while reducing herd size and land use. Early-stage selection thus enables more strategic herd management.

The feasibility of embryonic genotyping was made possible by the development of whole-genome amplification techniques. Embryo biopsies yield limited DNA — approximately 100 pg of DNA in 10 cells—far below the 200 ng required for standard commercial exome sequencing assays. The multiple displacement amplification (MDA) method overcomes this limitation, providing rapid and robust genome-wide amplification for exome sequencing in SNPChips [2].

The present protocol outlines a streamlined method for embryo biopsy that can be implemented in any standard embryo production laboratory without compromising post-transfer developmental rates [3]. Bovine embryos are produced in vitro and cultured to 154–168 h post-insemination (h.p.i.), up to the blastocyst or expanded blastocyst (preferable) stage. Individual embryos are transferred to drops of buffered medium (TALP) for sample collection using a microblade. Biopsies are placed into 0.2 mL tubes or cryovials and cryopreserved. Embryos are evaluated post-biopsy and directed toward transfer or cryopreservation.

Subsequently, biopsy samples are thawed and lysed to expose and briefly denaturate DNA for isothermal amplification using Phi29 DNA polymerase. MDA produces high-molecular-weight DNA with a low error rate, yielding micrograms of DNA from a few cells. Amplified DNA, contained within 20 μL, is then submitted for genotyping.

Validation of the amplification reaction is challenging, as quantification of non-purified amplification products by UV absorption yields inaccurate results due to the presence of unused primers in the final reaction mixture. The most reliable approach is to ensure that the biopsy is present in the initial reaction tube, as the cost and consistency of quantification and integrity analysis do not justify their use in routine practice. Nonetheless, in this document, we present tests using microchannel electrophoresis results to analyze DNA fragment integrity, purity, and size, as well as next-generation sequencing (NGS) call rates using the BovineSNP50 assay for selected samples.

2 Materials

2.1 Hand-Made Embryo Biopsy

Equipment:

- Stereoscope suitable for embryo handling.
- Heating stage.
- −70 °C freezer or nitrogen tank.

Disposables:

- Micropipettes and tips for 2, 10, 100, and 1000 μL.
- Blades for microsurgery—suggestion: Surgistar traditional ophthalmic scalpel, # 961501 or 962261.
- Culture treated surface plates (35, 60, and 100 mm).
- 0.2 mL conical tubes suitable for thermocyclers or cryovials (for nitrogen tank storage, in case −70 °C freezer is unavailable).
- Sterile gauze.

Reagents:

- Mineral oil tested for embryo culture.
- Buffered medium—suggestion: Tyrode's acid lactate pyruvate.
- 3% agarose gel prepared in 35 mm plate.

Additional resources:

- Fine lab pen.
- Recipient with ice or liquid nitrogen to temporarily store biopsies.

2.2 WGA

Equipment:

- Thermocycler.
- Simple centrifuge for sample spin.
- −70 °C freezer or nitrogen tank.

Disposables:

- Micropipettes and filtered tips for 2, 10, 20, 200, and 1000 μL.
- 0.2 mL conical tubes suitable for thermocyclers or cryovials (for nitrogen tank storage, in case −70 °C freezer is unavailable).

Reagents:

- Ultrapure decontaminated DNAse-free water.
- GenomiPhi Single Cell DNA Amplification Kit (Cytiva) components:

- Reaction buffer.
- Enzyme mix.
- Amplification mix.
- DTT solution.
- Lysis buffer.
- Neutralization buffer.

Additional Resources:

- Fine lab pen.
- Recipient with ice to temporarily store reagents and biopsies before and after processing.

3 Methods

Follow the appropriate safety procedures and wear personal protective equipment, always including gloves and clothes. Handle reagents and samples with appropriate care.

3.1 Manual Embryo Biopsy (Main Steps Are Illustrated in Fig. 1)

Embryo biopsy by manual microblade excision will be described here (*see* **Note 1**).

1. Prepare a biopsy plate by adding 5 μL droplets of buffered medium in a 60 mm plate. Cover with mineral oil and add more 5 μL per droplet. Identify each droplet with an indexed number that will contain embryo morphology and affiliation information in your lab form.
2. Select grade I blastocysts at 144–168 h post-insemination (h.p.i.) interval. Wash the embryos in 100 μL buffered medium. Place each of them into the identified droplets in the biopsy plate.
3. With the microblade, apply a slight pressure on the plate to scratch the surface. Roll the embryo to the mark with the inner cell mass to the opposite side of the tentative biopsy portion (Fig. 1, I).
4. Cut the tentative biopsy portion starting from the blade tip to the end (Fig. 1, II). Repeat the movement if the biopsy is not fully separated from the embryo.
5. Make sure the biopsy is not attached to the microblade prior to microblade removal from the droplet. If it is attached, proceed with caution to gently remove it, using the bottom of the dish and/or a 10 μL tip (*see* **Note 2**).

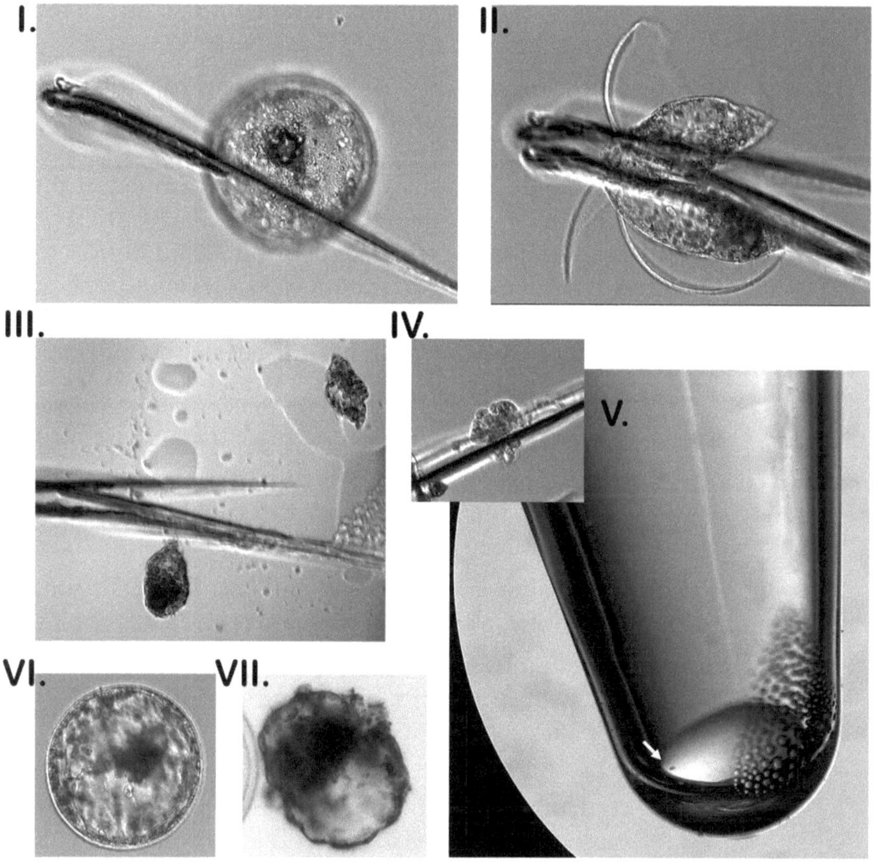

Fig. 1 Steps involved in manual embryo biopsy. I. Expanded blastocyst positioned on microblade scratch just before biopsy. II. Blastocyst sectioned just after biopsy. III. Separation of blastocyst containing inner cell mass (bottom) and biopsy (top). IV. A biopsy of 10–15 cells partially attached to a microblade scratch. V. Tube (0.2 mL) containing 1 μL droplet with the embryo biopsy (arrow) (stereoscope view). VI. Expanded blastocyst prior to embryo biopsy procedure. VII. Expanded blastocyst 2 h after the biopsy procedure

6. Clean the microblade by cutting agar gel and sterile gauze. Rinse it in buffered medium and proceed to the next biopsy (*see* **Note 3**).
7. When all embryos are processed (usually 20–30 per 60 mm plate), place the plate in the heating stage and identify the vials with all relevant information (e.g., oocyte donor, semen, embryo quality, embryo number, date).
8. Using a micropipette calibrated to 1 μL, collect an embryo biopsy from the plate and place it carefully in the bottom of the identified vial, make sure to produce a round droplet easily seen (this will help for biopsy processing). This step should be performed under a stereoscope to make sure the biopsy is transferred adequately (Fig. 1, V) (*see* **Note 4**).

9. Store into −70 °C ultra freezer (in 0.2 mL tubes) or liquid nitrogen tank (in cryovials) (*see* **Note 5**).
10. Biopsied embryos should be directed to vitrification or transferred to recipients, following protocols elsewhere [3, 4].

3.2 Biopsy Processing for WGA

Cell lysis and DNA denaturation are performed in the lysis step (step 1); WGA is achieved by adding Phi29 DNA Polymerase, random modified hexamers, nucleotides, and buffers to the denatured DNA—modified random hexamers nonspecifically bind to the DNA, and isothermal amplification proceeds (step 2). After amplification, the enzyme is heat-inactivated (step3) and the sample is ready for genotyping.

1. Collect vials with biopsy and place them on ice for defrosting.
2. If the biopsy sample is stored in a cryovial, identify a 0.2 mL tube with the information—the sample will be transferred to the new tube.
3. Prepare lysis buffer: For each sample, 1 μL DTT solution +9 μL lysis buffer. Vortex and spin down.
4. Prepare mix: For each sample, 11 μL reaction buffer and 4 μL water. Keep the mix on ice and wait to add the enzyme and amplification mix at step 2.
5. (Step 1—Cell lysis and DNA denaturation) Locate the drop containing embryo biopsy and include 1 μL lysis buffer. If the biopsy is in a cryovial, collect the 2 μL suspension and transfer to the identified 0.2 mL tube. Incubate at 65 °C for 10 min (*see* **Note 4**).
6. Add 1 μL neutralization buffer and gently pipette.
7. (Step 2—WGA) Add Phi29 enzyme (1 μL enzyme mix) and 1 μL amplification mix for each sample (*see* **Note 6**). Gently pipette the mix and add 17 μL to each sample tube. If necessary, spin the tubes to concentrate the liquid at the bottom. Incubate at 30 °C for 2 h.
8. (Step 3—Enzyme inactivation) Incubate at 65 °C for 10 min.
9. The tubes are ready to be sent to the genotyping service, preferably using dry ice. No WGA product quantification is performed routinely (*see* **Note 7**).
10. *See* **Note 8** for an optimization of all protocols to increase the number of samples per kit.
11. *See* **Note 9** for genotyping service options.

4 Notes

1. *Biopsy Protocol*

   Embryo biopsy can be performed either by micromanipulation using holding and biopsy pipettes or by manual microblade excision (described here). Where micromanipulators are available, biopsy via micropipette reduces embryo damage, with loss rates below 10%, compared to 20–30% for manual excision. Loss rates vary significantly with operator experience. Importantly, both methods yield similar pregnancy and delivery rates compared to non-biopsied embryos; differences are mainly seen in immediate post-biopsy recovery.

2. *Biopsy Collection*

   The ideal biopsy consists of a compact cell cluster. However, if cells disperse across the plate surface, careful pipetting and aspiration can help consolidate and collect the fragments. A pipette tip may assist gentle nudging but must be used carefully to avoid cell shearing or adhesion to the pipette.

   The transfer of the biopsy sample from the plate to the pipette tip, and then into the collection tube, should be carefully monitored under a stereoscope.

3. *DNA Contamination*

   Strict attention must be paid to removing any residual cumulus cells attached to the zona pellucida, as these can introduce maternal DNA contamination. To minimize cross-sample contamination, the blade should be cleaned between each biopsy, using agar gel, sterile gauze, and rinsing.

4. *DNA Integrity*

   Degenerated or extruded embryonic cells should be avoided due to the risk of DNA degradation. The Single Cell GenomiPhi DNA Amplification Kit is optimized for high-fidelity whole genome amplification from individual mammalian cells. Using apoptotic or degraded cells increases amplification bias.

   During the cell lysis and DNA denaturation step, incubation time and temperature must be carefully controlled to preserve DNA integrity.

5. *Biopsy Storage*

   If −70 °C storage is unavailable, samples can be cryopreserved in nitrogen tanks using cryogenic vials. In this case, 0.2 mL tubes should not be used due to the risk of explosion during thawing.

6. *Phi29 Enzyme Handling*

   Handle the enzyme with care and ensure the kit is used as promptly as possible within its validity period. It is recommended to collect a batch of samples and process them

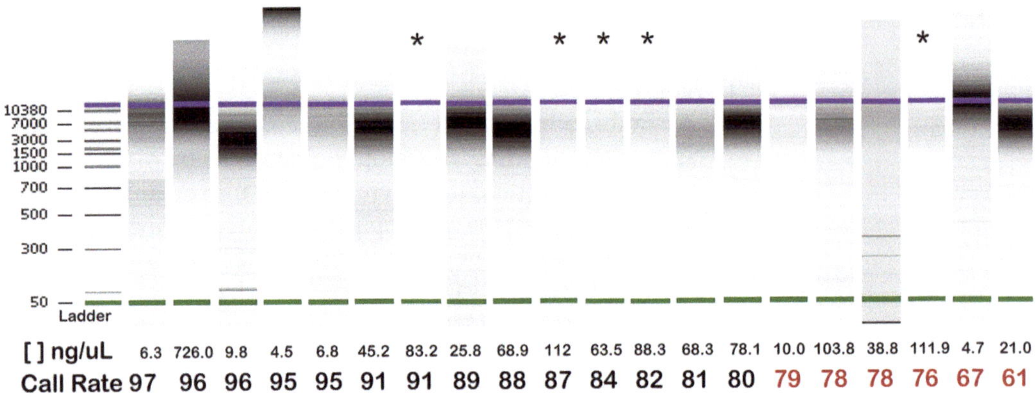

Fig. 2 Montage of gel-like images representing electropherogram plots of 20 biopsy samples after WGA. The Bioanalyzer DNA 12000 series was used. Ladder and main values are shown in the left gel. DNA concentration and call rate values after genotyping with Bovine 50 k SNP ChiP, Illumina, are shown. Asterisks indicate samples processed using 0.5 reaction volumes

together for efficiency. It is critical that the enzyme be stored properly. The enzyme xix should be stored at −80 °C. If the material will be consumed within 2 months, −20 °C storage may be used. The freezer must not be a frost-free unit.

7. *WGA Optimization for Increased Number of Samples Processed Per Kit*

   Alternative processing using half-reactions (marked with an asterisk in Fig. 2) was tested by initiating amplification with 0.5 μL sample droplets. This approach doubled the processed sample number without compromising the result quality in a limited pilot study. Figure 4 compares mean call rates among genomic DNA, full-reaction WGA samples, and half-reaction WGA samples. No significant differences were observed between full and half WGA reactions, although both showed the expected lower call rate values relative to genomic DNA due to amplification bias.

   Thus, *half-reaction optimization* offers a viable strategy for improving efficiency without significant loss of data quality.

8. *DNA Quantification and Integrity Following WGA*

   If validation is required, microchannel electrophoresis can be performed to assess DNA fragment integrity, purity, and size. We tested our samples using Agilent 2100 Bioanalyzer System (Bioanalyzer DNA 12000 Series) according to manufacturer instructions (https://www.agilent.com/cs/library/usermanuals/public/G2938-90024_DNA7500-12000_KG.pdf) in 20 WGA biopsy samples sent to the genotyping service. Also, call rates using the BovineSNP50 assay for genotyping were registered and compared to quantification.

In Fig. 2, a montage of all images in the same figure is presented, with the ladder on the left side. Each electropherogram plot presents a gel-like image, with DNA concentration and call rate values below.

All samples generated DNA smears with fragments predominantly between 1500 and 10,000 base pairs. Markers at 17,000 bp (purple) and 50 bp (green) are shown. Samples are ordered from highest to lowest call rate; however, no clear pattern differentiates high- and low-call rate samples. This is because fragment size and concentration alone do not predict genome coverage, which must be assessed through sequencing. Figure 3 demonstrates the nonsignificant correlation between DNA concentration and call rates.

Due to intrinsic limitations of whole-genome amplification, complete SNP coverage is not always feasible, and WGA samples typically achieve lower call rates compared to genomic DNA. As a result, breeding programs using WGA samples often use *genomic imputation*, and even low-call rate samples can yield reliable genetic value estimates depending on the imputation model.

9. *Genotyping Using Commercial Bead Chips*

Several commercial bead chips for microarray assays containing informative single-nucleotide polymorphisms distributed across the genome are currently available for bovine samples, and customized assays can also be performed [1]. This assay is usually performed by an external service provider. The user should define which SNP Chip is preferred to use,

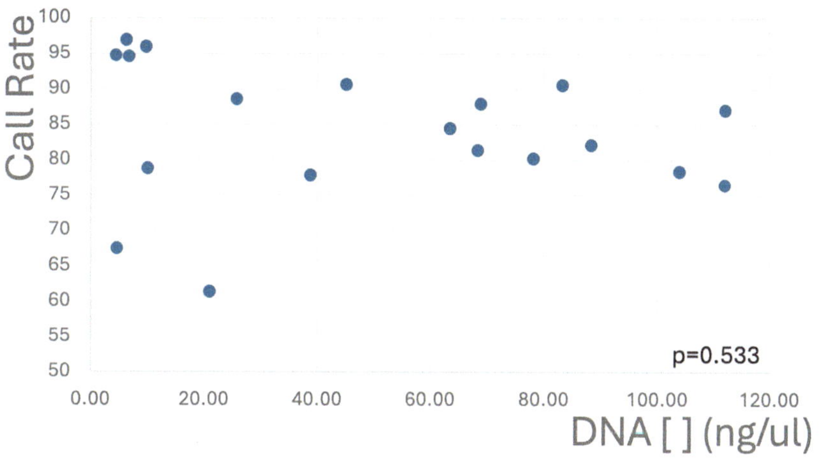

Fig. 3 Correlation between call rate values after genotyping with Bovine 50 k SNP ChiP, Illumina, and DNA concentration estimated by Bioanalyzer DNA 12000 series before genotyping

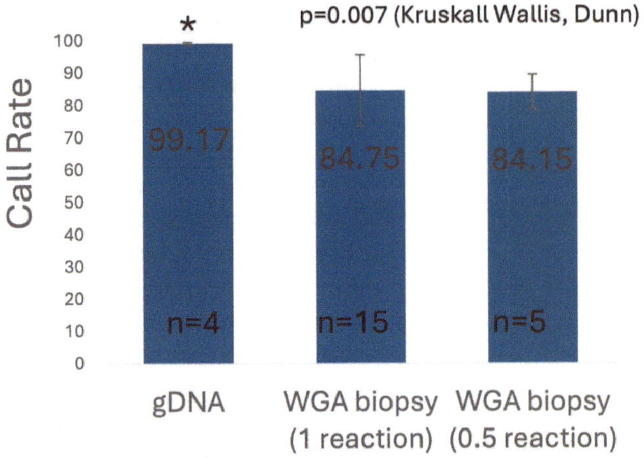

Fig. 4 Mean call rate values after genotyping with Bovine 50 k SNP ChiP, Illumina, in the same assay, from genomic DNA samples (gDNA) and biopsy WGA samples processed using 1 or 0.5 reaction

considering further required analysis. For breeding programs, it is crucial to arrange analysis before genotyping, since reference population and imputation methods would affect this decision, aiming for the best accuracy [2].

References

1. Oliveira CS, Camargo LSA, Silva MVGB et al (2023) Embryo biopsies for genomic selection in tropical dairy cattle. Anim Reprod 20(2): e20230064
2. Oliveira CS, Silva MVGB, Quintao CC et al (2023) Imputation accuracy for genomic selection using embryo biopsy samples in Gir. Reprod Biol 23(2):100765
3. Oliveira CS, Romano Quintão CC, Freitas C et al (2017) Post implantation development reveals that biopsy procedure can segregate "healthy" from "unhealthy" bovine embryos and prevent miscarriages. Anim Reprod Sci 184:51–58
4. Oliveira CS, Feuchard VLS, Freitas C et al (2020) In-straw warming protocol improves survival of vitrified embryos and allows direct transfer in cattle. Cryobiology 97:222–225

Chapter 8

Accelerating Genomics: Innovations in Analyzing Limited and Rare Samples

Ioanna Andreou, Markus Storbeck, and Jonathan Schaffer

Abstract

Single-cell genomic analysis offers insights into various biological fields but is challenged by the limited DNA from single cells. Whole genome amplification (WGA) techniques, such as PCR and multiple displacement amplification (MDA), facilitate the creation of high-quality next-generation sequencing (NGS) libraries. Advances in sequencing technologies and enrichment methods have reduced costs and enabled the analysis of a significant number of single cells. Effective cell isolation techniques maintain cell viability and integrity. This article presents a streamlined workflow that includes single-cell dispensing, genome amplification, and NGS library generation.

Key words WGA, MDA, NGS, Single cell genomics, Hybrid capture, PCR, Cell isolation

1 Introduction

Single-cell genomic analysis provides valuable insights across a wide range of research areas, such as developmental biology, tumor heterogeneity, and disease pathogenesis. However, performing single-cell genomic analysis using next-generation sequencing (NGS) has traditionally been challenging due to the limited amount of genomic DNA available from a single cell. Creating high-quality next-generation libraries typically requires more DNA than what is contained in a single cell. It is essential to amplify the material uniformly while avoiding the introduction of errors.

Whole genome amplification (WGA) is a field with continuous development and ongoing advancements. Various methods have been established, all relying on DNA polymerase synthesis and employing different priming strategies, including specific, degenerate, or hybrid primers [1]. WGA techniques can involve PCR amplification, primer extension, or multiple displacement amplification (MDA). Innovations in multiple displacement amplification (MDA) technologies allow for the unbiased amplification of small

Yafeng Ma (ed.), *Whole Genome Amplification: Methods and Protocols*, Methods in Molecular Biology, vol. 3003,
https://doi.org/10.1007/978-1-0716-5060-8_8,

DNA quantities [2] while maintaining their integrity. By yielding significant amounts of DNA, this amplification method supports the construction of high-complexity NGS libraries, which offer optimal coverage for precise variant calling [3].

Recent advancements in sequencing platforms and chemistries have significantly reduced the cost of whole genome sequencing (WGS). Enrichment of specific regions of interest can further reduce the sequencing output required, thereby lowering associated costs. Two primary enrichment methods are commonly used: PCR-based and hybrid-capture-based approaches. PCR-based methods are highly flexible and efficient but are typically designed to target specific loci known to be associated with disease. For inherited diseases of unknown genetic origin, a broader approach is needed to capture genetic variations not yet linked to any specific disease [4]. Hybrid-capture enrichment methods, on the other hand, require larger amounts of DNA, making exome enrichment from single cells feasible only after preamplification of the limited DNA available [5].

However, the accuracy and completeness of the amplification and quality of the generated sequencing data cannot be separated from the quality of the starting single cell. Before conducting a single-cell analysis, it is essential for scientists to identify and isolate the individual cells. The effectiveness of cell isolation technologies is typically assessed using three key parameters: efficiency or throughput (the number of cells isolated within a given time), purity (the proportion of target cells collected after separation), and recovery (the fraction of target cells retrieved after separation compared to the total number of target cells initially present in the sample) [6].

Current cell isolation methods can be broadly categorized into three groups based on the principles they employ. The first group relies on physical properties such as size, density, electrical charges, and deformability. Techniques in this category include density gradient centrifugation, membrane filtration, and microchip-based capture platforms, which have the advantage of enabling single-cell isolation without the need for labeling. The second group is based on the biological characteristics of cells and utilizes their protein expression properties to isolate individual cells [6]. The techniques include affinity-based methods, such as affinity solid matrix techniques using beads, plates, or fibers, fluorescence-activated cell sorting (FACS) or dispensing, and magnetic-activated cell sorting (MACS) [7]. The last method is limiting dilution, which will separate cells by generating serial dilutions.

Regardless of the technique used to isolate individual cells, it is crucial to preserve cell viability and maintain the integrity of the cell. This includes avoiding alterations in characteristics such as gene expression patterns or inducing apoptosis, which can compromise DNA integrity [8].

In this article, we provide tools for a streamlined workflow that allows studies of single cells. We have adapted single cell dispensing using the HANA cell dispenser (Bio-Techne) based on the physical surface characteristics of cultured cells to isolate single cells. We have amplified the complete genome of these single cells and generated whole genome and whole exome NGS libraries [5, 9, 10]. The libraries were qualified for adapter/adapter dimers and quantified in order to generate a library pool for suitable sequencing on NGS platforms.

2 Materials

Prepare all materials with the reagent provided in the respective listed kits. Store all reagents according to the instructions on the kit boxes and vials. Thaw components stored at −35 to −15 °C on ice. Always wear a suitable lab coat, disposable gloves, and protective goggles when handling chemicals. For further details, refer to the relevant safety data sheets (SDSs). Adhere strictly to waste disposal regulations when disposing of waste materials.

2.1 Cell Preparation and Counting

1. PBS (Gibco™ PBS, pH 7.2, cat. no. 12549079).
2. Trypan Blue Dye (Bio-Rad, cat. no. 1450021).
3. Cell Counting Slides (Bio-Rad, cat. no. 1450015).
4. Pipettes 1–10 μL, 20–200 μL, and 100–1000 μL and pipette tips (e.g., Thermo Fisher, Sarstedt).
5. TC20 Automated Cell Counter (Bio-Rad #14501102).
6. Microcentrifuge tubes, PCR-strips, or plates (e.g., Eppendorf, CORNING).
7. Microcentrifuge (e.g., Eppendorf).
8. Vortexer (e.g., VWR).
9. Ice.

2.2 Cell Sorting

1. PBS (Gibco™ PBS, pH 7.2, cat. no. 12549079).
2. Water OmniPur® Grade, Sterile Purified, WFI Quality, Cell Culture Tested (Merck, cat. no. 486505).
3. Hana Cell Sorter and Single Cell Dispenser (Bio-Techne, cat. no. NI004).
4. Single Cell Microfluidic Cartridge Kit (Bio-Techne, cat.no. NC003).
5. Microfluidic Cartridge Cleaning Kit (Bio-Techne, cat.no. NC301).
6. Pipettes 1–10 μL, 20–200 μL, and 100–1000 μL and pipette tips (e.g., Thermo Fisher, Sarstedt).
7. Microcentrifuge tubes, PCR-plates (e.g., Eppendorf, CORNING).

2.3 WGA

1. Isolated cells or purified DNA in <4 μL PBS sc.
2. PBS sc: PBS buffer specially treated to eliminate any DNA contamination.
3. TE buffer (10 mM Tris-HCl; 1 mM EDTA, pH 8.0).
4. QIAseq Single Cell DNA Library Kit UDI (QIAGEN, Cat no: 181703, 181705, or 181707):
   (a) PBS sc 1x.
   (b) Buffer DLB: This buffer is freeze-dried, resuspended in 500 μL of the included H2O sc to obtain DLB buffer pH > 13.
   (c) DTT, 1 M: Dilute DTT to 0.1 M with H_2O sc by mixing 5 μL DTT 1 M and 45 μL H2O sc. The dilution should always be freshly prepared.
   (d) Buffer D2: Mix 4.4 μL DDT 1 M and 48.4 μL Buffer DLB. This quantity is sufficient for 16 reactions plus an additional 10%. Adjust the amounts proportionally if a greater number of reactions is needed. Store residual Buffer D2 at −20 °C. Buffer D2 should not be stored for longer than 3 months.
   (e) Stop solution.
   (f) REPLI-g sc DNA Polymerase. Thaw the polymerase on ice to maintain its activity.
   (g) REPLI-g sc Dilution Buffer. Thaw at room temperature and then place on ice. This buffer may form a precipitate after thawing. The precipitate will dissolve by vortexing for ~10 s.
   (h) REPLI-g sc Advanced Oligo for human cells or DNA or REPLI-g sc Universal Oligo for all species. Thaw at room temperature and then place on ice.
   (i) H2O sc. Thaw at room temperature and then place on ice.
   (j) QIAseq Human Control DNA. Thaw and place the polymerase on ice.
5. Pipettes 1–10 μL, 20–200 μL, and 100–1000 μL and pipette tips (e.g., Thermo Fisher, Sarstedt).
6. Thermal cycler (e.g., Eppendorf, Bio-Rad).
7. Microcentrifuge tubes, PCR-strips, or plates (e.g., Eppendorf, CORNING).
8. Microcentrifuge (e.g., Eppendorf).
9. Vortexer (e.g., VWR).
10. Ice.

2.4 Whole Genome Amplification-DNA and Library Quality Control

1. Capillary electrophoresis device, e.g., QIAGEN QIAxcel®, Agilent® Bioanalyzer®, or similar to evaluate the DNA fragmentation profile (optional).
2. QIAxcel DNA High Sensitivity Kit (QIAGEN, cat. no. 929012).
3. Qubit 4 fluorometer (Thermo Fisher Scientific, cat. no. Q33238).
4. Qubit 1X dsDNA HS Assay-Kit (Thermo Fisher Scientific, cat. no. Q33230, Q33231).
5. Qubit™ Assay Tubes (500 tubes, cat. no. Q32856) or Qubit™ Flex Assay Tube Strips (cat. no. Q33252).

2.5 FX Library

1. QIAseq Single Cell DNA Library Kit UDI (Cat no: 181703, 181705, or 181707):
   (a) FX enzyme mix.
   (b) FX buffer.
   (c) FX enhancer.
   (d) DNA Ligase.
   (e) DNA Ligase buffer.
   (f) Plate with Unique Dual-Index (UDI) adapters compatible with Illumina sequencing platforms.
   (g) QIAseq Beads: Allow the mixture to equilibrate at room temperature for 30 min before use. Mix thoroughly to resuspend the beads, ensuring the solution is homogeneous and brown.
2. 80% Ethanol: Freshly prepare by mixing 40 mL Ethyl alcohol, Pure, ≥99.5%, ACS reagent (Sigma) with 10 mL Water OmniPur® Grade, Sterile Purified, WFI Quality, Cell Culture Tested (Merck, cat. no. 486505). This quantity is sufficient for 16 reactions plus an additional 25%.
3. Pipettes 1–10 μL, 20–200 μL, and 100–1000 μL and pipette tips (e.g., Thermo Fisher, Sarstedt).
4. Thermal cycler (e.g., Eppendorf, BIO-RAD).
5. Microcentrifuge tubes, PCR-strips, or plates (e.g., Eppendorf, CORNING).
6. Microcentrifuge (e.g., Eppendorf).
7. Vortexer (e.g., VWR).
8. Magnetic rack for 2 mL microcentrifuge tubes or PCR plates (Thermo Fisher DynaMag™-2 Magnet, cat no: 12321D, or Dynamo™-96 Magnet, cat no: 12027).
9. Ice.

2.6 Hybrid Capture

1. QIAseq Human Exome Kit (QIAGEN, cat. no. 333937, 333939).
   - (a) Human Exome Probe Set.
   - (b) One-4-All Blocking Oligos.
   - (c) One-4-All Blocking Solution.
   - (d) Fast Hybridization Solution.
   - (e) Vapor-Lock.
   - (f) Post Hybrid-Capture PCR Mix (2x).
   - (g) Primer Mix Illumina Library Amplification.
   - (h) RNase-free water.
   - (i) Streptavidin Binding Beads.
   - (j) Post-Capture Binding Buffer.
   - (k) Wash Buffer A.
   - (l) Wash Buffer B.
   - (m) Nuclease-free water.
2. QIAseq Beads (QIAGEN, cat. no. 333923, 333903).
3. 100% ethanol (ACS grade).
4. Buffer EB (QIAGEN, cat. no. 19086).
5. Pipettes 1–10 μL, 20–200 μL, and 100–1000 μL and pipette tips (e.g., Thermo Fisher, Sarstedt).
6. Thermal cycler (e.g., Eppendorf, BIO-RAD).
7. Microcentrifuge tubes, PCR-strips, or plates (e.g., Eppendorf, CORNING).
8. Microcentrifuge (e.g., Eppendorf).
9. Vortexer (e.g., VWR).
10. Magnetic racks for magnetic bead separation (e.g., Thermo Fisher Scientific DynaMag™ Magnet).
11. Two heating blocks for 1.5–2.0 mL tubes (e.g., Eppendorf).
12. An evaporation device (e.g., Eppendorf Concentrator Plus, cat. no. 305000100).

3 Methods

3.1 Cell Preparation and Counting

1. Transfer the detached cells or cell suspension from the bottle into a 2 mL microtube using a pipette and centrifuge at 300x *g* for 5 min (*see* **Note 1**).
2. Remove the supernatant and add 500 μL of PBS sc. Mix and centrifuge again.

3. Remove the supernatant and add 200 μL of PBS sc. Mix and place on ice.
4. Mix 10 μL of cell suspension with 10 μL of trypan blue. Gently pipette up and down 5–10 times to mix the cells and dye [11] (*see* **Note 2**).
5. Pipette 10 μL of the cell mix into the outer opening of one of the chambers of the counting slide.
6. Completely insert the slide into the slide slot of the TC20 instrument to start the counting procedure (*see* **Note 3**).
7. Remove the slide after counting is complete, and the total count in cell/mL, live count in cells/mL, and % of live cells are visible on the instrument screen. The dilution with trypan blue is accounted for 1 to 1 by the instrument.
8. Proceed with cell dispensation using a cell suspension with 90% or higher live cell viability.

3.2 Cell Dispensing

1. Carefully read the instructions supplied with the instrument (*see* **Note 4**).
2. Fill the bottle with PBS and place it in the corresponding position on the instrument. Close the bottle with the cap connected to the instrument's tubing.
3. Empty the waste bottle and close it with the cap connected to the instrument's tubing.
4. Switch on the instrument.
5. Switch on the PC of the instrument.
6. Attach the USB cable to the PC to connect the instrument to the PC.
7. Start the software and follow the step-by-step instructions provided by the software wizard. The wizard will guide you through all steps, from initialization to analysis, gating, dispensing, and finally, shutting down the instrument.
8. Begin instrument initialization with the cleaning cartridge in the cartridge holder and the holder closed. The pump will activate, and you will observe a drip of liquid from the waste tubing into the waste container. Initialization is complete when no bubbles are visible in the tubing and the liquid drip is steady.
9. Dilute the cell suspension in PBS to a concentration of 8000 cells/mL, then pipette 650 μL of the diluted cell suspension into a new cell cartridge.
10. Replace the cleaning cartridge with the cell cartridge in the cartridge holder. Follow the steps outlined by the software wizard to place waste collector on the stage and prime the cell cartridge (*see* **Note 5**).

11. Set the trigger for forward side scatter (FSC) at a level of 100, corresponding to 10 μm, which is the typical size of a mammalian cell. This will determine the threshold for a cell event. Start the cell analysis. The cell data will appear as dots on the plots.
12. After analyzing 50 cells, stop the analysis and set the gates for dispensing. Adjust the scale of the scatter plot to visualize all analyzed cells and draw a box around the cell population to be dispensed (Fig. 1). The analysis is complete, and cells can be dispensed.

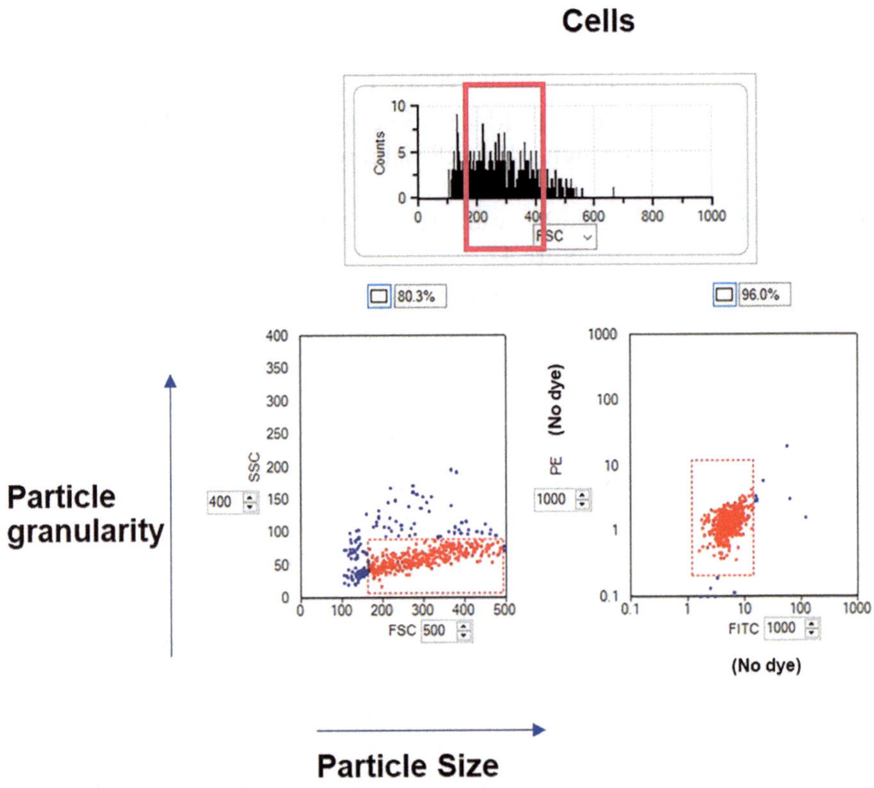

Fig. 1 Cell analysis and dispensing
The image shows scatter plots and a histogram used for analyzing cell data and identifying gated cell populations:
Top Histogram: Displays cell counts against forward scatter (FSC), highlighting a specific range and showing the distribution of values for the selected detection channel.
Bottom Left Scatter Plot: Shows side scatter (SSC) versus FSC, indicating particle granularity and size. Cells appear as blue points, while cells meeting the specified trigger threshold (dotted line box/gate) appear as red data points, representing a different cell population or subset. The cells included in the red gate represent 80.3% of the total cell population.
Bottom Right Scatter Plot: Displays PE dye versus FITC dye. Currently, no dyes have been used, so the values for these dyes are low. The dotted red box gates a well-defined cell population, which represents 96% of the cells.
This visualization helps in identifying and gating cell populations based on their scatter properties and dye intensities

13. Replace the waste collector with a PCR plate on the stage.
14. In the dispensing mode, select the plate type and the wells for cell dispensing. If starting with a dilution of 8000 cells/mL, the dispensing of 96 single cells will take approximately 5 min. The time may vary depending on the concentration.
15. After dispensing is complete, follow the instructions of the software wizard to clean and shut down the instrument.

3.3 Whole Genome Amplification

1. For whole genome amplification of the dispensed cells, use the QIAseq Single Cell DNA Library Kit UDI. This kit is designed for efficient and reliable amplification of DNA from single cells, ensuring high-quality library preparation for downstream applications.
2. Thaw H_2O sc, DTT, and REPLI-g sc dilution buffer at room temperature. Vortex and briefly centrifuge them. If the REPLI-g sc Reaction Buffer forms a precipitate after thawing, dissolve it by vortexing for 10 s.
3. Gently mix the REPLI-g sc Polymerase by tapping the tube. Spin down for a few seconds and place on ice until use.
4. Set the thermal cycler to 30 °C or enter the program listed in Table 1. If using a thermal cycler with a heated lid, set the lid temperature to 70 °C.
5. Add 500 μL H_2O SC to the DLB-tube to dissolve the DLB buffer. Mix thoroughly and centrifuge briefly (*see* **Note 6**).
6. Prepare sufficient denaturation Buffer D2 by adding the components as listed in Table 2 and for the total number of whole genome amplification reactions.

Table 1
Cycling conditions of MDA

Time	Temperature	Comment
2 h	30 °C	Multiple displacement amplification
3 min	65 °C	Enzyme inactivation
Hold	4 °C	Cooling of amplified material

Table 2
Preparation of D2 buffer

Component	Volume for 16 reactions (plus 10%)
DTT 1 M	4.4 μL
Buffer DLB	48.4 μL

Table 3
Preparation of DTT 1:10 dilution

Component	Volume for 16 reactions (plus 10%)
DTT 1 M	4.4 μL
H_2O sc	39.6 μL

Table 4
Cell Lysis and DNA denaturation protocol

Cell lysis and DNA denaturation		Denaturation of purified DNA	
Time	Temperature	Time	Temperature
10 min	65 °C	3 min	25 °C
1 min	4 °C	1 min	4 °C

The table represents the volumes needed for 16 reactions with 10% overfill to avoid any loss of material due to pipetting errors. If more samples are processed in parallel, upscale accordingly (*see* **Note 7**).

7. Dilute DTT with H_2O sc as described in Table 3. The table represents the volumes needed for 16 reactions with 10% overfill and can be upscaled accordingly if more samples are being processed in parallel (*see* **Note 8**).
8. Place 4 μL cell material or purified DNA (in PBS) into a microcentrifuge tube. If using less than 4 μL of cell material or DNA, add PBS sc to bring the volume to 4 μL (*see* **Note 9**).
9. Add 3 μL Buffer D2 to the 4 μL sample. Mix carefully by gently flicking the tube and centrifuge briefly (*see* **Note 10**).
10. Incubate the cell material with D2 buffer in the thermal cycler by running the protocol "Cell lysis and DNA denaturation" (Table 4) and immediately proceed with the next step and neutralization of lysate. If running control experiments with control QIAseq Human Control DNA, use the protocol "Denaturation of purified DNA" (*see* **Note 11**).
11. Add 3 μL stop solution. Mix carefully by flicking the tube and centrifuge briefly. Immediately place on ice (*see* **Note 12**).
12. Prepare REPLI-g master mix by adding components in the order listed in Table 5. Then, mix by pulse vortexing at low rpm and centrifuge briefly (*see* **Note 13**).
13. Add 40 μL master mix to each 10 μL denatured DNA (from **step 11**). Mix carefully by flicking the tube and centrifuge briefly.

Table 5
Preparation of REPLI-g master mix

Component	Volume per reaction	Volume for 16 reactions (plus 10%)
H_2O sc	6.5 μL	114.4 μL
REPLI-g sc dilution buffer	14.5 μL	255.2 μL
REPLI-g sc universal oligo or REPLI-g sc advanced oligo	14.5 μL	255.2 μL
DTT (1:10)	2.5 μL	44 μL
REPLI-g sc DNA polymerase	2 μL	35.2 μL

14. Place the tube in the thermal cycler and run the protocol "Cycling conditions of MDA" listed in Table 1.
15. After completion of amplification proceed with the amplified DNA quality control or store amplified DNA at 4 °C for short-term storage (2–4 weeks) o −20 °C for long-term storage (>4 weeks) (*see* **Note 14**).

3.4 Whole Genome Amplification-DNA and Library Quality Control

1. Before starting with the procedure, carefully read the Qubit™ 1X dsDNA HS Assay Kit user guide [12].
2. Equilibrate Qubit™ 1X dsDNA HS Working Solution (Component A), Qubit™ dsDNA BR Standard #1 (Component B), and Qubit™ dsDNA BR Standard #2 (Component C) for 15–30 min at room temperature (*see* **Note 15**).
3. Calibrate the Qubit™ Fluorometer using the Qubit™ 1X dsDNA HS Assay, which requires both standards, component B and component C. Mix 10 μL of each standard with 190 μL of the Qubit™ 1X dsDNA HS Working Solution. Vortex and spin down.
4. Prepare the WGA sample by diluting it 1–10 with H_2O sc. Mix 1 μL of each diluted WGA sample or 1 μL of undiluted library with 199 μL of the Qubit™ 1X dsDNA HS Working Solution to achieve a final volume of 200 μL (*see* **Note 16**). Vortex and spin down.
5. Allow all tubes standards, WGA sample, or libraries to incubate at room temperature for 2 min, then proceed to read standards and samples.
6. On the "Home" screen Qubit™ 4 Fluorometer, touch dsDNA, then select 1X dsDNA High Sensitivity (HS) assay icon, and select "Read standards" to proceed (*see* **Note 17**).
7. Insert the tube containing Standard #1 into the sample chamber, close the lid, and select "Read standard." The reading will

be completed in approximately 3 s. Remove Standard #1 and insert the tube containing Standard #2 into the sample chamber. Perform the reading for Standard #2 and remove the tube once the reading is complete. The result is displayed on the Read Standards screen (*see* **Note 18**).

8. To proceed with sample quantification, select "Next." Then, select the sample volume (1 μL) and units by using the + or— buttons on the wheel, or anywhere on the wheel itself.
9. Insert a sample tube into the sample chamber, close the lid, and select "Run samples." The reading will be complete in approximately 3 s. Remove the sample tube. Standards and samples are displayed on a graph with the results in a list below. View the results list by selecting the graph icon. The value listed is the concentration of the original sample (*see* **Note 18**).
10. To read more samples, select "Add samples" and repeat **step 9** until all samples have been read.

3.5 Library Generation

1. Thaw all kit components on ice. Once thawed, mix buffers gently by pulse vortexing to avoid localized concentrations. Briefly centrifuge reagents before use.
2. Equilibrate the QIAseq Beads to room temperature for 20–30 min before use and mix them until no visual clumps are present (*see* **Note 19**).
3. Program a thermal cycler according to Table 6. Set the thermocycler heated lid at 70 °C and start the program. When the temperature has reached 4 °C, pause the program.
4. Dilute the WGA-DNA to 50 ng/μL in H_2O sc and pipette 10 μL in a fresh PCR tube or plate placed on ice or a cooling block (*see* **Note 20**).
5. Prepare the FX Reaction Mix. Add the components of the FX Reaction Mix in the same order as stated in Table 7 and mix by pipetting. Vortex the buffer mix to ensure homogeneity before adding the FX Enzyme Mix. After adding the enzyme, pipette up and down, then keep the mix on ice. Scale up the FX Reaction Mix as needed for the number of samples.

Table 6
Cycling conditions for enzymatic fragmentation of MDA amplified material

Step	Temperature	gDNA (WGA)
1	4 °C	1 min
2	32 °C	15 min
3	65 °C	30 min
4	4 °C	Hold

Table 7
FX Reaction Mix

Component	Volume per reaction	Volume for 16 reactions (plus 10%)
FX buffer, 10x	5 μL	88 μL
H_2O SC	20 μL	352 μL
FX enhancer	5 μL	88 μL
FX enzyme mix	10 μL	176 μL

Table 8
Ligation master mix

Component	Volume per reaction	Volume for 16 reactions (plus 10%)
DNA ligase buffer, 5x	20 μL	352 μL
H_2O SC	15 μL	264 μL
DNA ligase	10 μL	176 μL

6. Pipette 40 μL FX Reaction Mix to each WGA-DNA sample on ice, gently vortex to mix, and spin down the PCR plate/tubes/strips.
7. Place the FX Reaction Mixes into the prechilled thermocycler (4 °C) and start the fragmentation program. Once the program is complete, transfer samples to ice and proceed immediately with adapter ligation.
8. Gently vortex and spin down the UDI adapter plate to prepare the ligation adapters. Remove the protective adapter plate lid, carefully pierce the foil seal with the pipette tip, and transfer 5 μL from one UDI adapter to each 50 μL sample from the previous protocol. Track the barcodes used for each sample (*see* **Note 21**).
9. Prepare the ligation master mix (per DNA sample) in a fresh 1.5 mL tube on ice according to Table 8. Mix well by gently pulse vortexing (*see* **Note 22**).
10. Add 45 μL of the ligation master mix to each sample from **step 8**. Close the tubes or seal the plate, mix well by pulse vortexing, spin down, and incubate at 20 °C for 15 min (*see* **Note 23**).
11. After ligation, remove unused adapters or adapter dimers by adding 60 μL (0.6x) of resuspended QIAseq Beads slurry to each ligated sample and mix well by pipetting or gently vortexing.
12. Incubate the mixture for 5 min at room temperature to bind the DNA to the beads.

13. Spin down to collect all liquid droplets, immobilize the beads on a magnetic stand for 2 min, and discard the supernatant.
14. Add 200 μL freshly prepared 80% ethanol to each pellet to wash the beads. Pellet the beads on the magnetic stand for 2 min. Control visually if the supernatant is clear, and carefully discard the supernatant without disturbing the bead pellet with the pipette tip.
15. Repeat the ethanol wash once more.
16. Incubate on the magnetic stand for 5–10 min to dry the beads (*see* **Note 24**).
17. Remove from the magnetic stand and vortex to resuspend the pellet in 52.5 μL H_2O sc and to elute the DNA. Pellet beads on the magnetic stand after 2 min of incubation at room temperature. Carefully transfer 50 μL of the clear supernatant to a new PCR plate/tube.
18. Perform a second cleanup using 55 μL (1.1x) QIAseq Beads.
19. Mix well and repeat **steps 12–16** to bind and wash unbound fragments.
20. Finally, elute each library by resuspending the bead pellet in 25 μL H_2O sc. Vortex to mix and wait for 2 min, then immobilize the beads on the magnetic stand. Carefully transfer 22.5 μL of the clear supernatant into a new PCR plate. Store purified libraries at −20 °C.
21. The hybrid capture protocol requires a minimum of 200 ng of blunt-ended libraries. The ligated UDI adapters have a typical Y-shape and require four cycles of library amplification to become blunt-ended.
22. Program a thermocycler according to Table 9.
23. For library amplification, prepare a reaction mix according to Table 10. Mix by pulse vortexing and spin down.
24. Place the samples in the cycler and start the cycling program as outlined in Table 9.

    After PCR amplification, store the library at −15 to −35 °C if not purified immediately.
25. Purify the amplified library with 1.0x QIAseq Beads purification by adding 50 μL resuspended QIAseq Beads slurry to each amplified library and mix well by pulse vortexing.
26. Incubate the mixture for 5 min at room temperature.
27. Spin down to collect remaining droplets, immobilize the beads on a magnetic stand for 2 min, and carefully discard the clear supernatant.
28. Wash the beads by adding 200 μL fresh 80% ethanol to each pellet. Pellet the beads on the magnetic stand for 2 min, then carefully discard the supernatant.

Table 9
Library amplification conditions

Time	Temperature	Number of cycles
2 min	98 °C	1
20 s 30 s 30 s	98 °C 60 °C 72 °C	4
1 min	72 °C	1

Table 10
Library amplification reaction mix

Component	Volume/reaction (μL)
HiFi PCR master mix, 2x	25
Illumina library Ampl. Primer mix (10 μM each)	1.5
Library DNA	20
RNase-free water	3.5
Total reaction volume	50

29. Repeat the wash **step 28** once for a total of 2 ethanol washes.
30. Incubate on the magnetic stand for 5–10 min or until the beads are dry (*see* **Note 24**).
31. Elute by resuspending in 26 μL H_2O SC, pH 8.0. Pellet beads on the magnetic stand. Carefully transfer 23.5 μL supernatant to a new low-binding tube.
32. Store purified libraries at −15 to −35 °C until ready for sequencing or further processing.
33. Quantify the library as outlined in Subheading 3.4: "Whole genome amplification-DNA and library quality control."
34. Assess the quality of the library using a capillary electrophoresis device, e.g., QIAxcel Connect (QIAGEN) or TapeStation instrument and Kits (Agilent) or other comparable method (*see* **Note 25**).

3.6 Hybrid Capture-Based Enrichment

1. After quantifying the individual libraries, equal amounts of each library (in ng) should be pooled into a single tube. Using the same amount per sample ensures balanced sequencing read depth across all multiplexed libraries. The maximum total input mass for the pooled library is 3.2 μg (*see* **Note 26**).
2. Transfer the calculated volumes for each library to a 1.5 mL LoBind tube. Use a separate LoBind tube for each hybrid capture reaction.

Table 11
Complement the library pool with probes and blocking reagents

Component	Volume per reaction
Pool of indexed libraries (from step 1, basic protocol 3)	Variable
Human exome probe set	4 μL
One-4-all blocking oligos	8 μL
One-4-all blocking solution	5 μL

Table 12
Conditions for capture hybridization

Time	Temperature	Number of cycles
1 min	95 °C	1
5 min	95 °C	1
30 min to overnight	60 °C	Hold

3. Thaw the Human Exome Probe Set, All-4-One Blocking Oligos, and All-4-One Blocking Solution on ice. Mix well by pulse vortex and spin down.
4. Add the Human Exome Probe Set and blocking reagents to each library pool as described in Table 11, and mix by pulse vortexing and pulse spin the tube.
5. Dry the library pool mixed with probes and blocking reagents using a SpeedVac system or a similar evaporator. Set the temperature to 60 °C and continue the process until all liquid has fully evaporated (*see* **Note 27**).
6. Place the dried library pool on ice and proceed with the hybridization capture protocol (**step** 7). Alternatively, the dried library pool can be stored at −30 to −15 °C for up to 7 days.
7. Thaw the Vapor-Lock reagent and equilibrate to room temperature.
8. Preheat the Fast Hybridization Solution to 65 °C in a heating block, and ensure that any precipitate is fully dissolved. Vortex the solution periodically during the incubation—this process may take 10 min or longer. Keep the solution at 65 °C until it is ready to be used. **Steps** 7 and **8** can be performed during the evaporation of the library pool to save time.
9. Program a thermocycler according to Table 12, and set the heated lid at 85 °C to prevent evaporation during the capture hybridization.

10. Start the program and pause it when the thermocycler block reaches 95 °C.
11. Vortex the Fast Hybridization Solution (maintained at 65 °C), and then immediately transfer 20 μL to the tube containing the dried, complemented library pool from **Step 6**. Vortex the tube for 5 s, and then incubate for 2 min at 65 °C using the same heating block used to preheat the solution (*see* **Note 28**).
12. Vortex the sample from **step 11** for 5 s and place it again at 65 °C for 2 more minutes. Repeat this step once more for a total of 3 times and then pulse-spin the tube.
13. Carefully transfer the entire hybridization mixture into a 0.2 mL PCR tube, avoiding the formation of bubbles. Pulse-spin briefly to collect the contents at the bottom of the tube.
14. Add 30 μL of Vapor-Lock reagent to the PCR tube. Seal the tube tightly and pulse-spin to ensure all liquid collects at the bottom. The Vapor-Lock reagent will form a protective layer above the hybridization mixture, preventing evaporation without interfering with the hybridization process.
15. Place the tube into the thermocycler and continue the paused program at 95 °C and run hybridization for 4 h or more (*see* **Note 29**).
16. At least 30 min before the hybridization is complete, begin the protocol "Bind Hybridized Targets to Streptavidin Beads." Keep the hybridization reaction in the thermocycler at the designated temperature until it is ready to be transferred for binding to the Streptavidin beads (*see* **Note 30**).
17. Preheat one of the two heating blocks for 1.5 mL tubes to 70 °C.
18. Preheat the second heating block for 1.5 mL tubes to 48 °C.
19. Inspect the Post-Capture Binding Buffer, Wash Buffer A, and Wash Buffer B for any precipitates before use. If precipitates are present, heat the buffers in the original bottles to 48 °C, and mix thoroughly until the precipitates are completely dissolved.
20. Equilibrate Streptavidin Binding Beads to room temperature for at least 30 min.
21. Aliquot 450 μL of Wash Buffer A per hybridization reaction, preheat to 70 °C, and maintain this temperature by keeping the buffer in the heating block.
22. Aliquot 700 μL of Wash Buffer B per hybridization reaction, preheat to 48 °C, and keep the buffer in the heat block.
23. Bring the Streptavidin Binding Beads to room temperature and homogenize by vortexing thoroughly.
24. Pipette 100 μL of the homogenized Streptavidin Binding Beads into a new 1.5 mL LoBind tube for each hybridization

reaction. Make sure to mix continuously during pipetting to prevent the beads from settling.

25. Wash the beads from storage buffer by adding 200 μL Post-Capture Binding Buffer to the beads. Immobilize the beads on a magnetic stand for 1 min or until the solution is clear, and discard the clear supernatant without disturbing the pellet.
26. Remove from the magnetic stand and repeat **steps 25** and **26** two more times for a total of 3 washing steps.
27. After removing the supernatant from the final wash, add 200 μL of Post-Capture Binding Buffer to the beads. Vortex the tube to homogenize the bead slurry and keep it at room temperature until use.
28. Once the hybridization reaction is complete, open the lid of the thermocycler but keep the tubes inside to maintain the temperature. Immediately transfer the entire reaction volume, including the Vapor-Lock, into the tube containing the bead slurry from **Step 27**. Mix thoroughly by pipetting (*see* **Note 31**).
29. Place the tube on a shaker or rotator and incubate at room temperature for 30 min, ensuring continuous gentle mixing to prevent the beads from settling. Alternatively, incubate at room temperature for 30 min, manually agitating the tube every 5 min.
30. Pulse-spin the tube to collect all liquid at the bottom. Place the tube on a magnetic stand for 1 min, or until the solution is clear. Carefully remove and discard the clear supernatant, including the Vapor-Lock reagent, without disturbing the pellet. Residual Vapor-Lock reagent may be present during wash steps but will not affect the final capture product.
31. Remove the tube from the magnetic stand and add 200 μL of preheated 70 °C Wash Buffer A. Mix thoroughly by pipetting and incubate for 5 min at 70 °C in the heating block.
32. Return the tube to the magnetic stand for 1 min or until the solution is clear. Remove and discard the supernatant without disturbing the pellet.
33. Remove the tube from the magnetic stand and repeat the wash: Add another 200 μL of preheated Wash Buffer A, mix by pipetting, and incubate for 5 min at 70 °C.
34. Pulse-spin the tube and transfer the entire volume, including beads, to a new 1.5 mL DNA LoBind tube. Place the tube on the magnetic stand for 1 min or until the solution is clear, then carefully remove and discard the supernatant.
35. Proceed with Wash Buffer B: Remove the tube from the magnetic stand, add 200 μL of preheated (48 °C) Wash Buffer B,

and mix by pipetting. Pulse-spin briefly, then incubate for 5 min at 48 °C in a second preheated heating block.

36. Place the tube on the magnetic stand for 1 min, then remove and discard the clear supernatant without disturbing the pellet.
37. Repeat **Steps 35** and **36** two more times for a total of three washes with Wash Buffer B.
38. After the final wash, use a 10 μL pipette to remove any residual supernatant. Proceed immediately to the next step without allowing the beads to dry.
39. Remove the tube from the magnetic stand and add 45 μL of RNase-free water. Homogenize by pipetting and store the bead slurry on ice. This is a safe stopping point. Split the slurry into two 22.5 μL aliquots and store at −20 °C until further use. Do not discard the beads, as the captured DNA remains bound to the Streptavidin beads.
40. Proceed with the "post-capture amplification" (*see* **Note 32**).
41. Before setting up the amplification reaction, equilibrate QIAseq Beads to room temperature for 20–30 min, and prepare a fresh 80% ethanol solution.
42. Thaw Post Hybrid-Capture PCR Mix and Primer Mix Illumina Library Amplification on ice and mix by pulse-vortexing to avoid any localized concentrations. Then, briefly spin down.
43. Program a thermocycler with a heated lid according to Table 13.
44. Prepare the post-capture amplification reaction mix on ice according to Table 14. Combine components in a 0.2 mL PCR tube or an 8-well PCR strip, depending on experiment format.
45. Transfer the PCR tube or strip to the thermocycler and start the program.
46. After PCR is complete, add 75 μL (1.5x) of resuspended QIAseq Beads to each 50 μL reaction and mix thoroughly by pipetting up and down.

Table 13
Post-capture amplification conditions

Time	Temperature	Number of cycles
2 min	98 °C	1
20 s 30 s 30 s	98 °C 60 °C 72 °C	7, or 8 (by less than 1000 ng of library pool)
30 min to overnight	60 °C	Hold

Table 14
Reaction mix for post-capture amplification

Component	Volume per reaction
Bead slurry with captured targets (from step 39, basic protocol 3)	22.5 μL
Post hybrid capture PCR mix	25 μL
Primer mix illumina library amplification	2.5 μL

47. Incubate the mixture for 5 min at room temperature, then pellet the beads on a magnetic stand and carefully discard the supernatant.
48. Wash the beads by adding 200 μL of 80% ethanol. Pellet the beads again on the magnetic stand and discard the supernatant.
49. Repeat the wash once more for a total of two ethanol washes. Remove as much residual ethanol as possible.
50. Incubate on the magnetic stand for 5–10 min or until the beads are dry (*see* **Note 24**).
51. Elute by resuspending the beads in 32 μL of 10 mM Tris-HCl, pH 8.0, or RNase-free water. Incubate for 2 min at room temperature.
52. Pellet the beads on the magnetic stand and carefully transfer 30 μL of the clear supernatant to a new tube. The purified, post-capture library can be safely stored at −20 °C in a DNA LoBind tube until ready for sequencing.
53. Assess the quality and size distribution of the library using a capillary electrophoresis device such as QIAGEN QIAxcel Advanced or Agilent Bioanalyzer 2100 and quantify using Qubit as described in Subheding 3.4: "Whole genome amplification-DNA and library quality control" (*see* **Note 33**).
54. The library is ready for sequencing. For guidance on library dilution and loading onto the sequencer, please refer to the user manual and recommendations provided by the sequencing platform providers. For exome sequencing, it is recommended to allocate approximately 25–50 million read pairs per sample.

4 Notes (Link to the Text)

1. Due to the varying culture and detaching conditions required for different cell types, this step is not detailed in the protocol. Please refer to the published individual growth and detaching conditions for each cell type.
2. Trypan blue is toxic to cells, and extended exposure can lead to inaccuracies when assessing live/dead cell ratios.

3. Read carefully the safety information and operating instructions of the TC20 instrument in the corresponding User Manual [12]. Do not remove the slide during counting. If the number of cells is outside the detectable range of the instrument, an "out of range" value will appear on the current count screen. Press the down arrow key to select "View Image," then press Enter to examine the cell image, determine whether the sample should be diluted or concentrated, and repeat the cell count.
4. Installation and training for the Hana cell dispenser will be conducted by the supplier of the instrument.
5. During analysis and dispensing, monitor the sample level in the cartridge through the window in the cartridge holder, and ensure the cartridge does not run dry.
6. Reconstituted Buffer DLB can be stored at −20 °C for up to 6 months. Longer storage will decrease pH and result in inefficient denaturation of the DNA Duplex.
7. If performing fewer reactions, store residual Buffer D2 at −20 °C for a maximum of 3 months.
8. DTT dilutions should be freshly prepared before use and should not be stored.
9. Directly sort cells into PCR tubes or plates designated for the MDA reaction and keep them on ice. Prefill these tubes or plates with the appropriate volume of PBS SC to achieve a final volume of 4 μL ({4 μL PBS}—{X μL dispensing volume}). After sorting, centrifuge the tubes or plates to ensure cell droplets settle at the bottom of the wells. Mix reactions containing single cells by gently flicking the tubes to prevent cell loss on the pipette tip. Avoid pipetting single cells after sorting and before MDA to prevent cell material from sticking to the tip, which can result in sample loss. Do not sort cells in D2 buffer. Depending on the sorting instrument and cell concentration, this procedure may take between 5 and 60 min. Excessive incubation of cells and DNA in this buffer will compromise DNA integrity.
10. Ensure complete mixing of D2 buffer with cell material. Insufficient mixing will result in insufficient cell lysis and DNA release, which is necessary for effective amplification.
11. Purified DNA has undergone an isolation procedure that reduces its integrity. Therefore, it is denatured under mild alkaline conditions and does not require heating to prevent further DNA degradation, which could affect amplification uniformity. In contrast, DNA in cells is high-molecular-weight DNA, and heating is necessary to lyse the cell and nuclear membranes to fully release the DNA.

12. Once the stop solution is added to the denatured material, the buffer is neutralized, and the DNA will start to renature. Place the denatured material on ice to slow down this process and promptly add the REPLI-g reaction mix.
13. Before adding REPLI-g sc DNA Polymerase, briefly vortex and centrifuge the mixture of buffer, H_2O sc, DTT, and oligos. This step helps to remove any localized concentrations of salts and DTT, which could affect enzyme activity. The master mix should be kept on ice and used immediately upon addition of REPLI-g sc DNA Polymerase. This mix can be prepared during the incubation step to lyse the cells to reduce the time between the addition of the stop solution and REPLI-g master mix.
14. WGA-DNA should be treated as high molecular weight genomic DNA. Typical DNA yields range from 20 to 40 μg per 50 μL reaction, depending on the amount and quality of the starting material. Optical density (OD) measurements tend to overestimate REPLI-g amplified DNA. Therefore, it is recommended to use colorimetric methods such as PicoGreen™ [5] or Qubit™ for quantification. Dilute the DNA before quantification by adding an equal volume of H_2Osc to dilute salts that might influence Qubit™ accuracy.
15. Qubit™ assays are temperature sensitive, and accuracy can be influenced by temperature fluctuations. To minimize this, insert all assay tubes into the Qubit™ Fluorometer only for the time it takes to measure the fluorescence. Avoid holding the assay tubes in your hand before reading, as this can warm the solution and affect the readings. Be aware that Qubit™ Fluorometers can raise the temperature of the assay solution significantly, even over a few minutes. Additionally, the Qubit™ 1X dsDNA HS Working Solution is sensitive to light, but when protected from light, the signal is stable for 3 h. The assay is recommended for initial sample concentrations ranging from 10 pg/μL to 100 ng/μL.
16. The concentrations of WGA and libraries differ. To use the same Qubit™ assay for quantification, it is recommended to adjust the concentrations of WGA and libraries to fit within the assay's quantitative range for consistent quantification.
17. If you have already performed a calibration for the selected assay, the instrument will prompt you to either read new standards or use the previous calibration to run samples. You can skip this step and proceed directly to sample quantification. However, to minimize variables that can affect performance, it is recommended to perform a new calibration for every new assay.

18. For information on interpreting the calibration results, refer to the Qubit™ 4 Fluorometer User Guide (Pub. No. MAN0017209), available for download at thermofisher.com/qubit.

19. Store QIAseq Beads at 4–8 °C and never freeze them. After completing the protocol, return the beads to 4–8 °C. Prolonged room temperature incubation can acidify the solution and affect cleanup effectiveness.

20. Limit WGA-DNA input to 5 μL in the FX reaction to maintain salt concentration and fragmentation efficiency. If WGA-DNA is low in concentration and a higher volume is needed to achieve 500 ng DNA input, purify the WGA using QIAseq beads before the FX reaction [9].

21. It is important to change the pipette tip after pipetting an adapter into each sample to avoid cross-contamination, which can cause index switching between DNA fragments of different samples and lead to data integrity issues. Unused adapters can be frozen until further use. The adapter plate is stable for a minimum of 10 freeze–thaw cycles. Use only one adapter per ligation reaction.

22. Ligase buffer has a high density. Pipette and vortex carefully to ensure complete mixing. Include visual control between vortexing pulses.

23. Do not use a thermocycler with a heated lid. Ensure that the heated lid is cooled at room temperature before placing the ligation reaction in the cycler. If that is not possible, keep the thermocycler lid open or use a heating block.

24. Dried bead pellets should not be shiny or cracked. Ensure all ethanol residues are removed to avoid inhibition of subsequent reactions. Avoid over-drying, as it makes resuspension difficult.

25. The average library fragment size is approximately 400–450 bp (Fig. 2a), while adapter dimers appear in the 120–145 bp range. The median library fragment size can be used to calculate the library molarity based on the concentration measured by the Qubit™ assay. Although trace shapes may vary slightly between instruments, as shown in Fig. 3, where the same library was run on different platforms, these variations are minimal and do not significantly impact quantification.

26. To calculate the required volume (μL) of each amplified library, divide the desired mass by the library concentration. For example, for a library concentration of 20 ng/μL, 15 μL is needed to contribute 300 ng to the pool. When multiplexing 10 libraries at 300 ng each, the total pool will be 3 μg, staying within the allowed input range.

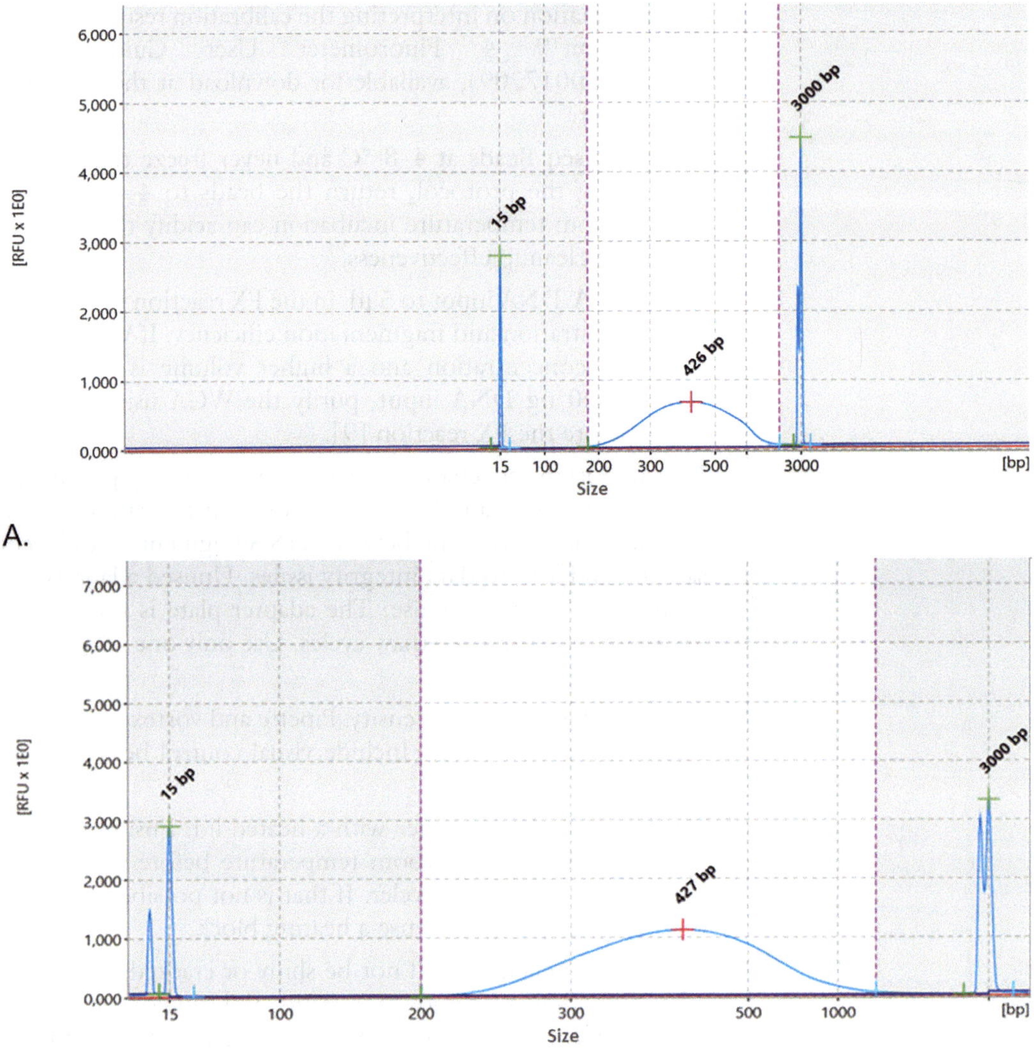

Fig. 2 Library fragment size distribution before and after the hybrid capture procedure as obtained on QIAxcel Connect System. (**a**) 0.75 ng of library was analyzed using QIAxcel DNA High Sensitivity Kit. The library is free of adapter and adapter dimer and has the typical shape of a QIAseq Single Cell DNA library with a median size of approx. 430 bp. (**b**) The library after the hybrid capture procedure was similarly analyzed using QIAxcel DNA High Sensitivity Kit. As expected, the size distribution has not altered after the hybrid capture procedure

27. The time required for complete evaporation will vary depending on the evaporation instrument used and the final volume of the library pool combined with hybrid capture probes and blocking reagents. As a general guideline, 200 μL of liquid typically requires 2–2.5 h to fully evaporate.
28. The Fast Hybridization Solution is highly viscous. It is important to pipette slowly to ensure accurate volume transfer. Do not allow the solution to cool down during handling. During

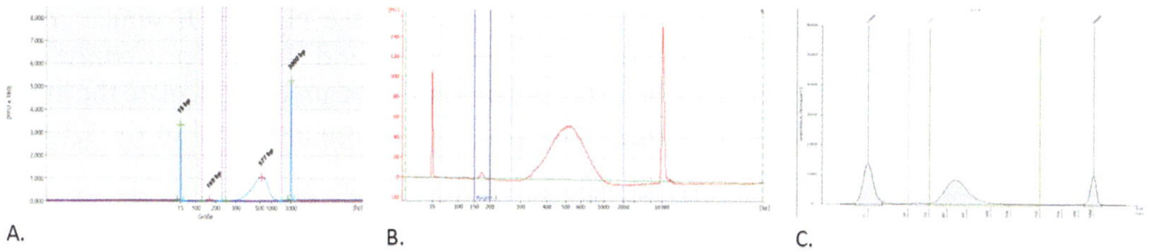

Fig. 3 Comparison of electrophoretic traces. (**a**) Electrophoretic trace of a library analyzed using the QIAxcel system (QIAGEN). The same library was also analyzed using the Agilent BioAnalyzer (**b**) and Agilent TapeStation (**c**)

this incubation, a white precipitate may form. This does not affect the hybridization procedure.

29. During the hybridization step, QIAseq Exome Probes bind specifically to target sequences within the library fragments present in the pool. The hybridization duration can be adjusted based on experimental needs, starting at 30 min and extending up to overnight. Incubating for 1 h or longer may improve capture performance, but durations exceeding 4 h will not significantly enhance hybrid capture performance. Overnight incubation can be considered for greater flexibility in experiment planning.

30. Do not allow the hybridization reaction to cool below 60 °C after the program is complete. In the subsequent steps, biotin-labeled Human Exome Probes and their hybridized target sequences will be captured using Streptavidin beads. Nonspecific fragments will be removed through a series of wash steps. To ensure a seamless workflow, begin preparing the Streptavidin Binding Beads and wash buffers at least 30 min before the end of the hybridization. For shorter hybridization durations (e.g., 1 h), start preparing the beads and buffers immediately after initiating the hybridization reaction.

31. It is important not to allow the hybridization reaction to cool below 60 °C before transferring to the Streptavidin Binding Beads. A rapid transfer directly from the thermocycler is crucial to minimize off-target binding.

32. In the procedure of "post-capture amplification," captured targets bound to Streptavidin Binding Beads will be amplified using QIAGEN's Post Hybrid-Capture PCR Mix. The amplified library will then be purified and prepared for QC and sequencing. The libraries have dedicated adapters for sequencing on Illumina® instruments, but with the appropriate conversion chemistry [13], they can be sequenced on other platforms, e.g., Aviti (Element Biosciences).

33. The expected size distribution (see Fig. 3b) is similar to the initial libraries used for the hybrid capture procedure. The median library fragment size can be used to calculate the library molarity based on the concentration measured by the Qubit™ assay.

References

1. Blainey P (2013) The future is now: single-cell genomics of bacteria and archaea. FEMS Microbiol Rev 37(3):407–427. https://doi.org/10.1111/1574-6976.12015
2. de Bourcy CF, De Vlaminck I, Kanbar JN, Wang J, Gawad C, Quake SR (2014) A quantitative comparison of single-cell whole genome amplification methods. PLoS One 9:e105585. https://doi.org/10.1371/journal.pone.0105585
3. Hou Y, Wu K, Shi X, Li F, Song L, Wu H, Dean M, Li G, Tsang S, Jiang R, Zhang X, Li B, Liu G, Bedekar N, Lu N, Xie G, Liang H, Chang L, Wang T, Chen J, Wang J (2015) Comparison of variations detection between whole-genome amplification methods used in single-cell resequencing. GigaScience 4:37. https://doi.org/10.1186/s13742-015-0068-3
4. Samorodnitsky E, Jewell BM, Hagopian R, Miya J, Wing MR, Lyon E, Damodaran S, Bhatt D, Reeser JW, Datta J, Roychowdhury S (2015) Evaluation of hybridization capture versus amplicon-based methods for whole-exome sequencing. Hum Mutat 36(9):903–914. https://doi.org/10.1002/humu.22825
5. Andreou I, Storbeck M, Hahn P, Rulli S, Lader E (2024) Exome sequencing starting from single cells. Curr Protoc 4(9):e70017. https://doi.org/10.1002/cpz1.70017
6. Hu P, Zhang W, Xin H, Deng G (2016) Single cell isolation and analysis. Front Cell Dev Biol 4(116). https://doi.org/10.3389/fcell.2016.00116
7. Dainiak MB, Kumar A, Galaev IY, Mattiasson B (2007) Methods in cell separations. Adv Biochem Eng Biotechnol 106:1–18. https://doi.org/10.1007/10_2007_069
8. Box A, DeLay M, Tighe S, Chittur SV, Bergeron A, Cochran M, Lopez P, Meyer EM, Saluk A, Thornton S, Brundage K (2020) Evaluating the effects of cell sorting on gene expression. J Biomol Tech 31(3):100–111. https://doi.org/10.7171/jbt.20-3103-004
9. QIAseq single cell dna library kit with unique dual indexes handbook. QIAseq single cell dna library kit with unique dual indexes handbook – QIAGEN
10. QIAseq human exome handbook. QIAseq human exome handbook – QIAGEN
11. Instruction manual TC20 automated cell counter. https://www.bio-rad.com/sites/default/files/webroot/web/pdf/lsr/literature/10024423.pdf
12. Qubit™ 1X dsDNA HS Assay Kits User Guide Qubit™ 1X dsDNA HS Assay Kits User Guide (MAN0017455 Rev C.0)
13. Cloudbreak™ Sequencing Kits AVITI Cloudbreak Sequencing Kits | Element Biosciences

Chapter 9

Sweet Sorghum Microspores' Fractionalization, Individual Isolation, and Whole-Genome Amplification

Chandana Kurapati, Mustapha Ọláwuni, Svitlana Lymanska, and Ahmad N. Aziz

Abstract

Sweet sorghum is one of the important crops that has been widely reported to be recalcitrant to genetic manipulation endeavors. It can serve as biofuel, food, and food per production systems toward energy sources as well as human and animal sustenance. To deploy biotechnology tools in advancement of this crop of such agronomic importance, genetic and molecular investigation for understanding is necessary. Underdeveloped cell walls are a significant physiochemical state for microspores per their developmental stages, which makes it ideal for DNA isolation, being readily lysed as natural protoplasts. After harvesting panicles from the plants grown in controlled greenhouse conditions those are separated into sections per developmental stages is the first step. During micro-gametogenesis the unicellular microspore divides asymmetrically to subsequently give rise to a mature pollen grain with a vegetative and generative cell, while the absence of the participation of the anther wall in pollen formation makes it possible to investigate these developmental events directly. Thereby, from distal to basal end along panicle length, the five stages—mid-binucleate, early-binucleate, late-uninucleate, mid-uninucleate, and early-uninucleate microspores—can be yielded toward fractionalization. Whole-genome amplification is ideally achieved through individually isolated microspores with underdeveloped exine from anthers obtained after aseptic handling of spikelets using 75% ethanol and 1% sodium hypochlorite as sterilizing agents. The free-floating microspores, thus separated from the surrounding anther wall tissues, yield single gametophyte-based high-quality genomic DNAs. Efficient breeding of sweet sorghum through genetic tools can be achieved through free microspore release after aseptic isolation and whole-genome amplification.

Key words Genomic DNAs, Micromanipulation, Sorghum pollen, Gametophyte stages

1 Introduction

Crop improvement via genetic transformation requires suitable recipient tissue for gene transfer, which is usually followed by regeneration of transformed plants, generally requiring a period in tissue culture. Sweet sorghum has been the subject of genetic and genomic studies for almost a century [1] but is a recalcitrant genus for tissue culture and selection of regenerable cell clumps from

Yafeng Ma (ed.), *Whole Genome Amplification: Methods and Protocols*, Methods in Molecular Biology, vol. 3003,
https://doi.org/10.1007/978-1-0716-5060-8_9,

single transformed cells. Emani et al. [2] indicated that among cereals, sorghum has been the most recalcitrant crop species for genetic modifications and, therefore, only a few reports are available on its successful transformation. DNA marker-based analyses of individual gametes have several downstream uses in plant breeding and genetic improvement applications [3, 4]. Investigating the dynamics of male gametophyte maturation, revealing the physiological and developmental processes, has proven to be challenging for the plant science community. Separation of five pollen developmental stages (early uninucleate, mid uninucleate, late uninucleate, early binucleate, and mid binucleate) from a single panicle with high purity is therefore needed toward the acquisition of reliable and reproducible biological data. The rate-zonal centrifugation [5] is one type of density step gradient technique that allows the complete separation of components in a mixture, based on their size and weight (mass), and these protocols are directly applicable to capture and identify microspore developmental stages. Such a separation procedure, which is based on discontinuous Percoll gradients, enables the purification of immature pollen per individual developmental stages without using a special and expensive cell sorting apparatus. Partitioning of panicle into five distinct portions (basal, post-basal, median, pre-distal, and distal) along its length, as well as that per color intensity (pale green to green) and size of spikelets within, facilitates the yield of the distinct pollen developmental stages as natural protoplasts. Natural pollen competition for fertilization leads male gametophyte selection per reproduction merit, which has implications for female-biased plant population ratios [6, 7]. In addition, gametophytes being responsive to culture and pollen germination approaches represent only a fraction of the total microspore population, which is available for DNA amplification causing bias in subsequent genetic analyses [8]. Also, the control of the sporophyte over the gametophyte has implications in plant evolution [6, 7], and this mechanism may be better understood by analyzing unbiased samples representing the entire microspore population. Plant gametes can be a convenient source of DNA for genetic analysis as only limited storage space is required for their use as direct meiotic segregation products. Production of unreduced gametes in plants can be through first or second division restitution [9], and isolated pollen analyses can reveal the source. Immature pollen grains represent natural haploids and, through DNA marker-based analyses can be used for genetic studies [4]; however, only 60–70% of germinated pollens and a few androgenic responsive microspores can be available for subsequent DNA analyses [3]. Early unicellular to mid-binucleate microspores are essentially free gametophytes as protoplast [10, 11], which can be individually captured for DNA isolation and sequencing after observing sorghum panicle characteristics past flag leaf deployment [12]. The process of single-cell DNA sequencing (sc DNA-seq) can be more challenging due to the limited amount of DNA; however, multiple displacement amplification (MDA) allows for whole-genome amplification (WGA) while generating high genome coverage (>90%) with a low (~10–7) false positive rate [13]. Notably, this study focuses on panicle gradation, microspores fractionalization, and WGA assessment per free individual gametophytes' DNA isolation in sweet sorghum.

2 Materials

The greenhouse and laboratory suppliers can be substituted as per the use of equivalent or comparable products. All reagents were prepared using ultrapure (18 MΩ.cm at 25 °C) and autoclave water.

2.1 Plant Materials

1. Seeds of two cultivars of *S. bicolor*, "Dale" (Plant Introduction [PI] number: 651495) and "Topper76-6" (PI number: 583832), both of which have been promoted for biofuel use in Tennessee by Delta Biorenewables (Memphis, Tennessee, USA), were obtained from the United States Department of Agriculture–Agricultural Research Service's Plant Genetic Resources Conservation Unit (Griffin, Georgia, USA).
2. Two additional sorghum cultivars, "Dasht Local" and "Achi Turi," were recommended for this study by the Agriculture Research Institute (Quetta, Balochistan, Pakistan) due to their value for biofuels, and their certified seeds were provided by Dr. M. Y. Khan Barozai (University of Balochistan, Quetta, Pakistan) as per the Pakistan–US Science and Technology Cooperation Program (international joint research) funded through the National Academies of Sciences, Engineering, and Medicine.
3. The fifth sweet sorghum variety "Sugar Drip" was obtained from Kentucky Farmers Co-Operative (Kentucky, USA).
4. 4.5–5 L nursery pots.
5. Miracle-Gro Potting Mix (Scotts Company, Marysville, Ohio, USA).
6. Systemic insecticide Bonide (Oriskany, New York, USA).
7. Rain-drip automatic watering systems (Fresno, California, USA).

2.2 Surface Sterilization

1. To make 70% ethanol, add 70.07 mL of 99.9% isopropyl alcohol (DUDA energy, Decatur, Alabama, USA) to 29.93 mL of ultrapure water to make volume up to 100 mL.

2. To prepare 100 mL of 1% sodium hypochlorite solution, mix 13.25 mL of 7.55% bleach (Clorox, Oakland, California, USA) with 86.75 mL of ultrapure water.
3. Plastic containers or magenta boxes with lids.

2.3 Fractionalization of Different Developmental Microspores

1. To prepare a liter of 1 M D-mannitol solution, 182 g of D-mannitol is dissolved in ultrapure water, and the volume is made up to 1000 mL.
2. Dissolve 18.2 g of D-mannitol in distilled water and make the volume up to 1000 mL to prepare 0.1 M D-mannitol solution or use dilution of 1 M D-mannitol solution.
3. Stock isotonic Percoll solution is prepared by mixing 900 mL of 100% Percoll containing 10% and 100 ml of 1 M D-mannitol solution per liter.
4. Prepare different Percoll gradients in 15 mL tubes by using the dilution chart described in Table 1.
5. Laminar hood and dissecting supplies (scalpel, tweezers, petri dishes, alcohol lamp, tool holder, etc.) to retrieve anthers from spikelets.

2.4 Microspores' Isolation, Counting, and Viability Test

1. Mini pellet mixer with cordless mortar (VWR Avantor, Radnor, Pennsylvania, USA) to release free microspores from anthers.
2. Dissolve 0.5 mg of DAPI in 1 mL of DMF and store the solution at −20 °C.
3. Nikon inverted microscope (Inverted Diaphot-TMD; Nikon, Tokyo, Japan).
4. Micromanipulator (Narishige International, Tokyo, Japan).
5. CellTram Microinjector (Eppendorf, Hamburg, Germany).
6. PCR tubes, 0.2 mL.
7. EVE™ Plus cell counter (NanoEnTek, Waltham, Massachusetts, USA).

Table 1
Discontinuous Percoll density gradient solutions prepared for the separation of microspores in 15 mL centrifuge (Thermo Fisher Scientific, Waltham, Massachusetts, USA) tubes

Stock isotonic Percoll (ml)	0.1 M D-mannitol (ml)	Gradient layer concentration[2] (vol/vol)	Total volume (ml)
2	18	10%	20
4	16	20%	20
6	14	30%	20
8	12	40%	20

2.5 Whole-Genome Amplification of Single-Gametophyte DNA

1. 6% (w/v) sorbitol solution is prepared by dissolving 6 g of D-sorbitol (Sigma-Aldrich or equivalent) in 100 mL of ultrapure water.
2. Phosphate-buffered saline (PBS) is prepared by dissolving 8 g of NaCl, 0.2 g of KCl, 1.44 g of Na_2HPO_4, and 0.24 g of KH_2PO_4 reagents in approximately 800 mL of ultrapure water and adjusting the pH to 7.4 for a final volume of 1 liter.
3. WGA kit known as REPLI-g Advanced DNA Single Cell Kit (Qiagen, Hidden, Germany).
4. Thermo cycler (Bio-Rad, Hercules, California, USA).
5. NanoDrop One spectrophotometer (Thermo Fisher Scientific).

2.6 Gel Electrophoresis

1. For the WGA products' separation, a horizontal agarose gel electrophoresis system (Fisher Scientific, Pittsburgh, PA, USA) is used.
2. Agarose gel (1.5% in Tris-Borate-EDTA buffer) analyses are also required to check DNA quality and amplification of the samples.
3. The WGA-based DNA products are loaded on gel using 6X loading dye (Fisher Scientific, Carlsbad, USA).
4. Sizes of the DNA fragments stained by 5X Gel Red (Biotium Inc., Hayward, California, USA) are determined by comparing with a DNA Mass™ Ladder (Invitrogen™ Life Technologies, Carlsbad, CA, USA) after electrophoresis in TBE (Tris-Borate-EDTA) buffer.
5. Gels are visualized/photographed over a UV transilluminator using an AlphaImager 2000 System (Alpha Innotech, San Leandro, CA, USA).
6. A personal computer with Windows operating system (Microsoft Corporation, Redmond, WA, USA) is attached to the Nikon D5300 camera with camera Pro2™ software (Nikon) for recording gel analysis data.

3 Methods

3.1 Growth and Treatment of Sorghum Plants

1. The seeds are germinated at room temperature in petri dishes, and then the plants are embedded in 4.5 L nursery pots containing Miracle-Gro Potting Mix (Scotts Company, Marysville, Ohio, USA).
2. For insect control, a half tablespoon of the systemic insecticide Bonide (Oriskany, New York, USA) is added per container following the manufacturer's application and formulation protocol.

3. Ten to 15 plants of each cultivar are maintained at 30 ± 3 °C during the day and 24 ± 3 °C at night in a greenhouse (located: latitude 36.12′N, longitude 86.89′W).
4. The pots are watered using rain drip automatic irrigating systems (Fresno, California, USA) with a setting of 8 h cycles.
5. The emerging panicles are scrutinized daily at the beginning of the booting stage, immediately after the emergence of the flag leaf.
6. Sheaths covering the immature inflorescence containing the developing pollen grains are gently peeled open to observe the physical appearance of the panicles.
7. Panicles are harvested by snipping the flag leaf stem two internodes beneath the crown of the plant, while recording the planting date, plant height, flag leaf length, and days since flag leaf deployment.
8. After collecting, the panicles are stored in resealable polypropylene bags at 4 °C in a refrigerator until use.

3.2 Surface Sterilization of Spikelets

1. Harvested panicles in the refrigerator are retrieved, while ensuring that the microspore developmental stages are optimal for subsequent use.
2. Panicles along the length are partitioned into five equal and distinct portions called basal, post-basal, median, pre-distal, and distal, using scissors and a ruler.
3. Minimize the tissue by removing excess parts when individually retrieving spikelets per five distinct portions (basal, post basal, median, pre-distal, and distal) of the panicle prior to surface sterilization.
4. Thereby, the excised spikelets per the above five classifications are collected in separate sterile containers or magenta boxes for the surface sterilization process, which removes contaminants such as bacteria, fungi, etc., from the surface of the explants.
5. Completely dip the spikelets in 70% ethanol (harsh disinfectant step) for 30 s in a sterile container.
6. By using a lid to retain small spikelets in a container, the 70% ethanol is replaced with 1% sodium hypochlorite solution.
7. Completely dip the spikelets in 1% sodium hypochlorite (mild disinfectant) for 15 min, while shaking the containers is to be continued.
8. Discard the 1% sodium hypochlorite using the container lid as a barrier to rinse the spikelet with sterile autoclaved water (tissue culture grade/ultrapure), at least 2–3 times (mildest) to completely remove disinfectant residues.
9. Plant explants should remain in sterilized condition for subsequent handling.

3.3 Microspore Isolation

1. The homogenization technique releases free microspores in mass while eliminating an amount of somatic anther wall and spikelet tissues.
2. Aseptic anther isolation method can be performed using a scalpel and tweezers to dissect from the spikelets, while care is used to avoid any damage to anthers.

3.3.1 Homogenization

1. Homogenization involves gently crushing surface-sterilized spikelets from five distinct portions (basal, post basal, median, pre-distal, and distal) of the panicle in 0.1 M D-mannitol solution while using an autoclaved mortar and pestle to release microspores.
2. The obtained microspore suspension is then filtered through the bleached/graded 40 cheese cloth to retain the spikelet and anther wall residues, and the pass-through is preserved for subsequent experimental procedures, including individual isolation.

3.3.2 Pellet Mixing of Anthers

1. In the pellet mixing method, sterilize spikelets per five panicle portions are collected in sterile petri plates and excised with the help of an autoclaved scalpel and forceps.
2. After excision of spikelets, approximately 50–60 anthers (obtained from 15 to 20 spikelets) are collected in 1.5 mL Eppendorf tubes filled with 300 μL of 0.1 M D-mannitol/androgenic media, depending on the subsequent experimental pathway.
3. To release microspores from anthers, the suspension in 0.1 M D-mannitol solution is agitated by using a cordless mortar and mini pestle (VWR Avantor, Radnor, Pennsylvania, USA) for 35 s.
4. This mechanical agitation breaks the anther tissues in solution and releases free microspores. Care is observed to balance mixing time per gametophyte viability while breaking anther tissue. Individual gametophyte isolation is possible from this suspension, or when applicable, these microspores can be used for tissue culture.

3.4 Gametophyte Viability Test

1. Use a 5 μL fraction of microspore suspension obtained from either the homogenization or pellet mixing procedure for gametophyte viability test.
2. Gametophyte viability can be assessed by mixing equal volumes (5 μL each) of microspore suspension and Trypan Blue (Sigma-Aldrich, Burlington, Massachusetts, USA) to observe gametophytes under a microscope.

3. Viability and cell counting can be conducted by analyzing the above mixture using the EVE™ Plus cell counter (NanoEntek, Waltham, Massachusetts, USA).

3.5 Fractionalization of Microspores at Different Stages

1. The larger gametophytes are present at the distal end (mid-binucleate) and the smaller ones are found closer to basal end (early uninucleate) of the sorghum panicles.
2. Layer the 2 mL of heaviest gradient at the bottom of the tube, followed by increments (Percoll density lowering) gradients in five different separate 15 mL Falcon tubes. Load 2 mL of each Percoll gradient slowly without agitation of each layer by slanting the Falcon tubes at a 45° angle when pipetting.
3. Gradient percentages are larger for microspores at the distal end and pre-distal portions than those from the basal and post-basal portions of the panicle. The microspore suspension in 0.1 M D-mannitol is diluted while maintaining cell density when being layered as 2 mL suspension in falcon tubes.
4. Centrifuge the tubes with Percoll gradients along with top layered microspores at 900 g (speed acceleration 5, speed braking 9) at 4 °C for 60 mins.
5. Visual confirmation of microspore interface bands at different gradients is observed for each portion of the panicle.
6. Performed an assessment of microspore fractions' purity or viability by using DAPI.

3.6 Individual Microspore Isolation

1. The aseptically excised anthers from sterilized spikelets using forceps and a scalpel can be directly crushed in 2 μL of 6% sorbitol on a glass slide if a microspore suspension prepared in 0.1 M D-mannitol is not used.
2. Pulled glass capillary with both open ends is attached to a pipette/electrode holder (Narishige International USA, Inc., East Meadow, NY, U.S.A), which is then used for aspiration through a cell Tram microinjector (Eppendorf, North America, Hauppauge, New York).
3. The microinjector is mounted on an ultrafine move Marzhauser (HS-6) micromanipulator (ALA Scientific Instruments, Inc., Westbury, NY, USA) to collect individual microspores. A 20-inch high platform (ALA Scientific Instruments, Inc., Westbury, NY, USA) for positioning the manipulator over the microscope is needed.
4. Care is taken to aspirate individual microspores with minimal suspension solution. The individual microspores are then transferred to 0.2 mL PCR tubes, which is then stored at −20 °C until use.

3.7 Whole-Genome Amplification of Single-Gametophyte DNA

1. Each tube (0.2 mL) with an individually isolated gametophyte is then centrifuged at 5000 rpm for 5 min, after phosphate-buffer saline or REPLI-g Advanced sc (single cell) storage buffer (Qiagen) is added to the tube to make the final volume of 4 μL.
2. Next, 3 μL of reconstituted denaturation buffer (D2) from the REPLI-g Sc kit (Qiagen) is added to the microcentrifuge tube before a 10-min incubation at room temperature.
3. Afterward, 3 μL stop solution (Qiagen) is added, and then the WGA reaction is performed.
4. Lysed gametophytes are then mixed by flicking the tubes and briefly centrifuged before being placed on ice.
5. A master mix is prepared by mixing (with back-and-forth pipetting) 9 mL of H_2O, 2 μL REPLI-g Sc DNA polymerase and 29 μL REPLI-g Advanced Sc reaction buffer.
6. This 40 μL of master mix is added to the tube containing 10 μL of denatured gametophyte samples individually.
7. The final 50 μL reaction volume is then centrifuged and incubated at 30 °C for 2 h in a thermos cycler (Bio-Rad, Hercules, California, USA) followed by a 3-min inactivation of the Phi 29 polymerase at 65 °C for later sink at 10 °C.
8. The resulting MDA products are then quantified using a NanoDrop One spectrophotometer (Thermo Fisher Scientific) before being subjected to gel electrophoresis.

3.8 WGA-Based Gel Electrophoresis

1. The WGA products (10 μL of the reaction mixtures) are separated by electrophoresis (3.5 Vcm—1 for 3 h) on 1.5% agarose gels stained with 5X Gel Red in 0.5X TBE (0.045 M Tris, 0.045 M boric acid, 0.001 M EDTA) buffer.
2. The gel is photographed over a UV transilluminator to visualize the 5X Gel Red stain using an AlphaImager 2000 System (Alpha Innotech, San Leandro, CA, USA).
3. Sizes of the WGA-induced DNA fragments is determined by comparing with a DNA Mass™ Ladder (Invitrogen™ Life Technologies, Carlsbad, CA, USA).
4. A personal computer with Windows operating system and camera Pro2™ software is used to control the gel and photographic exposure conditions while filtering UV light after monochrome photograph with D5300, camera Pro™ software (Nikon) scored the DNA profiles and saved the data as high resolution (*.TIF) picture.

4 Notes

1. By following surface sterilization protocol, the standardization issues for exploring the time of disinfectants are encountered, which may cause reduced viability of microspores over time.
2. The viability of microspores depends on the duration of pellet mixing over time due to the mechanical effects. Limited mixing time would leave the anther tissue intact, while extended mixing would reduce microspore viability.
3. Take more care during the Percoll gradient formation. The pressure on the pipette dispense button should be as low as possible (the flow rate should be less than 0.5 s during the 30 s dispense).
4. Discontinuous Percoll gradients must be prepared at least 1 h before the separation so that a uniform diffusion of particles is provided in each interphase. Formed Percoll gradients should not be stored longer than 12 h.
5. Fractions of developing microspores formed (as a distant band) after the centrifugation in gradients is collected with a syringe and a needle.
6. Gametophyte sizes ranged from 48.82 μm to 34.78 μm in the distal portion, 46.25 μm to 33.34 μm in the pre-distal portion, 43.95 μm to 32.86 μm in the median portion, 43.13 μm to 29.06 μm in the post-basal portion, and 32.32 μm to 25.25 μm in the basal portion of the panicles across various sweet sorghum varieties.
7. Gradient percentages ranged from 25% to 21.25% for the distal portion, 22.5% to 20% for the pre-distal portion, 21.25% to 16.25% for the median portion, 16.25% to 12.5% for the post-basal portion, and 16.25% to 12.5% for the basal portion from the panicles across various sweet sorghum varieties.
8. The WGA yielded DNA fragments from individual sweet sorghum microspores of similar size to the sheared DNAs obtained from any plant genomic sample.
9. The MDA procedure resulted in ample DNA in large fragments, which should be useful for downstream analyses.

Acknowledgments

1. The authors acknowledge the National Institute of Food and Agriculture for research grant funding and the Tennessee State University, College of Agriculture, for providing research facilities and graduate assistantships.

2. The authors extend sincere gratitude to Dr. Aron Felts and Mr. Dilovan Yahya, who, as graduate research assistants, provided their valuable contributions throughout the microspores' harvesting and individual isolations.

Conflict of Interest Statement *This study was funded by* the National Institute of Food and Agriculture, grant number TENX-2131-GFSHPP.Ahmad N. Aziz *has received a research grant from* the National Institute of Food and Agriculture.*The authors have no conflicts of interest to declare that are relevant to the content of this chapter.*

References

1. Smith CW, Frederiksen RA (eds) (2000) Sorghum: origin, history, technology, and production. Wiley
2. Emani C, Sunilkumar G, Rathore KS (2002) Transgene silencing and reactivation in sorghum. Plant Sci 162(2):181–192. https://doi.org/10.1016/S0168-9452(01)00559-3
3. Aziz AN, Sauve R (2008) DNA mapping of *Echinacea purpurea* via individual pollen DNA fingerprinting. Mol Breed 21(2):227–232
4. Aziz AN, Yakubu AM, Hamal SS (2017) Genotyping of individually isolated gametes from *Gossypium hirsutum* chromosome substitution lines. Single-Cell Biology 6(3):169
5. Dupl'áková N, Dobrev PI, Ren˅ák D, Honys D (2016) Rapid separation of *Arabidopsis* male gametophyte developmental stages using a Percoll gradient. Nat Protoc 11(10):2097–2112. https://doi.org/10.1038/nprot.2016.107
6. Lora J, Hormaza JI, Herrero M (2016) The diversity of the pollen tube pathway in plants: toward an increasing control by the sporophyte. Front Plant Sci 7:107
7. Delph LF (2019) Pollen competition is the mechanism underlying a variety of evolutionary phenomena in dioecious plants. New Phytol 224(3):1075–1079
8. Tai GCC, Seabrook JAE, Aziz AN (2000) Linkage analysis of anther-derived monoploids showing distorted segregation of molecular markers. Theor Appl Genet 101:126–130
9. Crespel L, Gudin S (2003) Evidence for the production of unreduced gametes by tetraploid *Rosa hybrida* L. Euphytica 133:65–69
10. Touraev A, Forster B, Jain SM (eds) (2009) Advances in haploid production in higher plants. Springer. https://doi.org/10.1007/978-1-4020-8873-5
11. Yamamoto Y, Nishimura M, Hara-Nishimura I, Noguchi T (2003) Behavior of vacuoles during microspore and pollen development in *Arabidopsis thaliana*. Plant Cell Physiol 44(11):1192–1201
12. Felts AM, Pokharel B, Yahya D, Aziz AN (2022) Whole-genome DNA amplifications from individually isolated sweet sorghum microspores. Applications in Plant Sciences 10(6):e11501. https://doi.org/10.1002/aps3.11501
13. Wang Y, Navin NE (2015) Advances and applications of single-cell sequencing technologies. Mol Cell 58(4):598–609

Chapter 10

Assessing Amplification Quality and Bias in MDA Methods Through Comparative Analysis of Short-Read Sequencing

E. D. Lozano-Escobar, V. Mateo-Cáceres, C. Mayoral-Campos, and M. Redrejo-Rodríguez

Abstract

Although high-throughput sequencing methods have greatly improved over the last few years, direct sequencing remains unfeasible when DNA quantity or quality is limited. In such instances, various whole genome or metagenome amplification (WGA) techniques can generate sufficient DNA for multiple analyses, albeit with some amplification bias. Competent WGA analysis is typically evaluated by sequence coverage, assessed through two key parameters: depth, referring to the number of reads containing each nucleotide, and breadth, indicating the proportion of nucleotides in the consensus sequence relative to the original sequence length at the obtained depth. Adequate coverage is essential for detailed genomic analysis and the detection of population variants, copy number variations (CNVs), and structural variants (SVs). This chapter outlines a pipeline for analyzing Illumina sequencing data of amplified samples compared to non-amplified samples to assess the performance of various WGA methods, starting from raw sequences.

Key words Multiple displacement amplification, Whole genome amplification, piPolB, piMDA, Φ29 DNA polymerase

1 Introduction

Nowadays, direct sequencing of DNA samples can be performed without prior amplification—reducing bias and improving data reliability. However, this is not always possible when the DNA quantity or quality is limited. Thus, sequencing of the original raw samples remains experimental or restricted to targeted testing in some instances [1–3]. Furthermore, factors such as sample origin and characteristics, genome isolation procedures, and sequencing platforms can also introduce error, bias, and underrepresentation of genomic sequences present in low quantities [4–6].

Authors Lozano-Escobar, V. Mateo-Cáceres, and C. Mayoral-Campos contributed equally to this work and alphabetically ordered.

Yafeng Ma (ed.), *Whole Genome Amplification: Methods and Protocols*, Methods in Molecular Biology, vol. 3003, https://doi.org/10.1007/978-1-0716-5060-8_10,

To ensure DNA sample availability for sequencing, whole (meta)genome amplification (WGA), which includes PCR-based approaches, multiple displacement amplification (MDA), and various derived methods, enables efficient amplification of minute nucleic acid samples. MDA allows faithful and reliable amplification of (meta)genomic DNA without prior knowledge of the target sequence and has been utilized for almost 25 years, even at the single-cell level (scWGA). Additionally, isothermal amplification protocols simplify the usage at the point of care and support various applications beyond genomic analysis. Most MDA protocols use hexamer random primers (RP) and the DNA polymerase from *Bacillus* virus Φ29 (Φ29DNAP), whose high processivity grants the generation of very long amplicons that facilitate high coverage and detection of single-nucleotide polymorphisms (SNP) [7]. However, MDA has some drawbacks, such as the generation of primer-related artifacts, chimeric DNA sequences, or biased amplification of sequences with extreme GC content, affecting uniform coverage and sensitivity in detecting minor alleles or underrepresented sequences. Furthermore, the high processivity of Φ29DNAP can lead to overamplification of circular DNA molecules, like plasmids or small circular genomes, limiting its use in certain metagenomic applications, such as metaviromic analysis. Consequently, several new tools and methods have been developed to enhance the accuracy of WGA, including thermostable Φ29DNAP variants [8, 9] and adapted procedures that waive the usage of random primers [10–12]. The effectiveness of each procedure may depend on the sample nature, making it challenging to identify the most suitable protocol. Therefore, a detailed assessment of new MDA alternatives involving samples with diverse quality, composition, and isolation sources/methods is required. In our previous work [11], we used a mock metagenome sample with a wide range of GC content and assessed the impact of different MDA protocols on coverage, GC bias, de novo assembly, and binning, as well as other comparative analyses.

2 Method Outline and Availability

This method describes a ten-step procedure for comparing diverse MDA protocols of whole metagenome amplification, represented in Fig. 1. As an example, we used a number of amplification experiments from a small (~18 MB) mock-metagenome [11]. The sample was generated mixing equal amounts of genomic DNA from reference laboratory strains of *Escherichia coli JM83* (NCBI reference genome NC_000913.3; GC = 50.79), *Bacillus subtilis 110NA* (NC_000964.3; GC = 43.51), *Pseudomonas aeruginosa PAO1* (NC_002516.2; GC = 66.37), and *Kocuria rhizophila ATCC 9341* (NC_010617.1; GC = 70.56). Reference

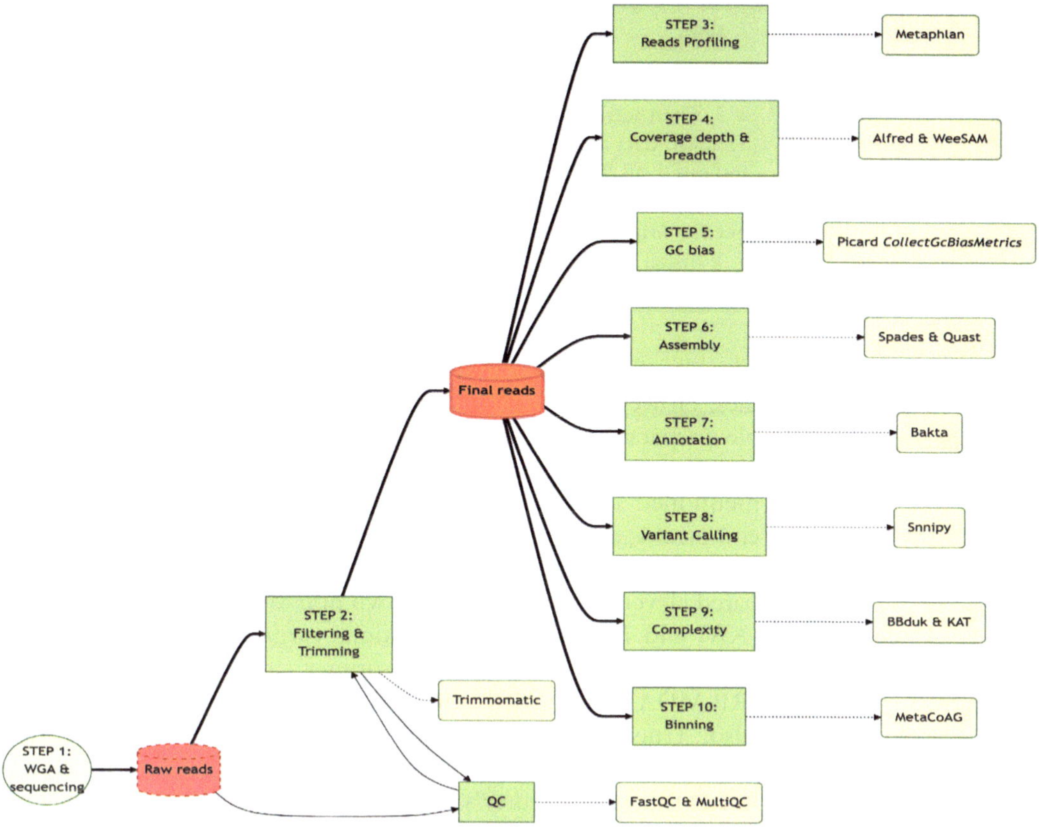

Fig. 1 Workflow of data analysis

genomes for amplification assessment were selected by BLASTN searches using the two longest contigs of each genome in the non-amplified sample (NA). Raw sequencing data have been deposited in the e-cienciaDatos repository with https://doi.org/10.21950/HCNDGF, and the original analysis is available in a GitHub repository as well as a Markdown report that details the statistical analysis and plots. This chapter provides an overview of the tools that can be used for a comparative, detailed analysis of MDA samples, including recent updates and example commands.

Note

We use exclusively Open-Source software that can be run on Linux or MacOS, as well as Windows under WSL (Windows Subsystem for Linux). Most of the packages can be installed seamlessly from Conda (see https://docs.conda.io/projects/conda/en/latest/user-guide/tasks/manage-environments.html). To avoid version incompatibility, we suggest creating one or more dedicated Conda environments for this analysis. All provided links were accessed at the preparation of this manuscript on March 10, 2025.

(continued)

> **Note**
>
> # 1. Create a new environment named 'wga' conda create --name wga
> # 2. Then activate env to use it conda activate wga
> Detailed data analysis and plotting are carried out in R. Unless otherwise indicated, all R packages can be installed from the CRAN repository as follows. We provide here some examples to illustrate and discuss the results. See the original repository for full details.
> # Install packages install.packages("ggplot2") # Load library (ggplot2)

2.1 Step 1: Whole Genome Amplification Reactions and Sequencing

The bioinformatics analysis pipeline starts from the raw short-read files. However, there are important considerations regarding sample preparation and sequencing. First, it is critical to select the appropriate methods according to the nature of your sample. As mentioned earlier, several factors, such as GC content, sample complexity, and the expected presence of circular plasmids or viral genomes, may affect the amplification bias. In our case, we compared eleven samples, spanning two non-amplified control samples (NA) and nine DNA amplifications carried out by six different protocols. First, a random-primer hexamers-based MDA (RP-MDA) was carried out in duplicate with REPLIg (Qiagen). Also, TruePrime (4basebio) was used as a control of primer-less MDA (PrimPol-MDA). In both cases, amplification protocols followed the manufacturer's instructions, including the routine alkaline denaturation step. Then, we used recombinant *E. coli* primer-independent PolB (piPolB), a replicative-like DNA polymerase encoded by mobile elements dubbed pipolins and endowed with DNA polymerase and 3′–5′ proofreading exonuclease activities as well as DNA priming capacity [13], to develop new MDA methods. Initially, piPolB was assayed as a single enzyme for primer-free DNA amplification in a single-step MDA protocol (piPolB-MDA) or an alternative protocol with a previous alkaline denaturation step (piPolB-MDA + D). Additionally, a new method based on the combination of piPolB and Φ29 DNA polymerases was developed. Two alternative protocols were used: a single-step primer-independent MDA (piMDA, also duplicated samples) and a modification with a previous alkaline denaturation of input DNA (piMDA+D, duplicated).

Sequencing facilities usually conduct routine sample QC tests on MDA reaction products before library preparation. These are often PCR-based tests that sometimes fail due to the accumulation of inorganic pyrophosphate during the reaction. Therefore, it is recommended to purify the DNA samples beforehand. Any quick procedure, such as DNA clean-up kits or traditional ethanol precipitation, can be used. In our experience, although ethanol-precipitated DNA is very convenient for shipping, remnants of ethanol may hinder DNA solubilization and thus interfere with library preparation. Regarding the preparation of sequencing libraries, it is highly recommended to use PCR-free methods, such as the TruSeq DNA PCR-Free Low Throughput Library Prep kit (Illumina) or equivalent kits to avoid PCR-derived artifacts or bias. Also consider the expected complexity of your sample when selecting the sequencing protocol. Ideally, a 200–300x depth, with a minimum 100x depth, is required for a detailed bias assessment. In our example, for a minimal metagenome of ~18 MB, a total length of ~2000–4000 MB would be needed, split into R1 and R2 reads if having paired-end reads.

2.2 Step 2: Raw Data Quality Check and Trimming

Once you receive the sequencing results, it is advisable to perform a quick preliminary check of your raw data, which can be quickly performed using SeqKit, giving an idea of the total length of your sequences.

```
# Option '-a' indicates 'all' stats and '-T' formats the output
as a table
# Option '-j' indicates core for parallelizing (in this example
12 cores)
seqkit stats -j 12 *.fastq.gz -a -T
# Alternative: Combined with 'csvtk' allow markdown format
conda install -c conda-forge csvtk
seqkit stats -j 12 *.fastq.gz -a -T | csvtk csv2md -t
```

Detailed quality check (QC), filtering, and trimming can be performed with FastQC and Trimmomatic. We used FastQC (v 0.12.1) for each sample, and QC reports can be aggregated with MultiQC (v 1.27.1), before and after filtering low-quality sequences and trimming out adapters with Trimmomatic (v. 0.39). Standard trimming parameters can be used for comparison between samples as long as all the samples are processed with the same parameters. However, note that improved metagenome binning for the extraction of metagenome-assembled genomes (MAGs, *see* **step 10**) might require parameters optimization, which is beyond the scope of this protocol.

```
# Install in Conda if needed

#A bash script is recommended to automate the process, like
that in https://github.com/lakhujanivijay/I-Shell-Do-This/
tree/master/automation_of_trimmomatic

# Code example for non-amplified sample (Ctrl_R1.fastq.gz;
Ctrl_R1.fastq.gz)
fastqc Ctrl_R1.fastq.gz -o QC_raw
fastqc Ctrl_R2.fastq.gz -o QC_raw
#Alternatively, for multiple samples:
fastqc *.fastq.gz -o QC_raw
```

```
# Merge reports with MultiQC
multiqc ./QC_raw
# Filtering and trimming reads
trimmomatic PE \
 -threads 12 \
  -phred33 Ctrl_R1.fastq.gz Ctrl_R2.fastq.gz trimmed/
Ctrl_R1_paired.fastq.gz trimmed/Ctrl_R1_unpaired.fastq.gz
trimmed/Ctrl_R2_paired.fastq.gz trimmed/Ctrl_R2_unpaired.
fastq.gz \
  ILLUMINACLIP:~/Trimmomatic-0.39/adapters/TruSeq3-PE.fa:2:
30:10:2:keepBothReads \
 LEADING:3 \
 TRAILING:3 \
 SLIDINGWINDOW:4:20 \
 MINLEN:35

#QC and MultiQC to the trimmed reads.
fastqc trimmed/*.fastq.gz -o QC_trimmed
multiqc ./QC_trimmed
```

Tip

MultiQC is a very handy tool for merging reports from diverse applications. To obtain publishable reports, you can tailor the report, including your customized title, comments, or even a logo, among other options. For instance, if you want to rename your files to match the experimental design, you can use the —replace-names option and provide a TSV table with the new names. Additionally, since January 2025, it can also contain an AI-generated summary of your data (options —ai and —ai-summary-full).

2.3 Step 3: Metagenome Profiling

Profiling metagenomes against databases allows for the detection and quantification of microorganisms, independently of the assembly and even at low abundances where assembling is not possible. Moreover, this procedure enables the identification of possible contaminant genetic material as well as DNA amplification artifacts, such as primer-derived spurious amplification, also referred to as template-independent DNA amplification (TIDA) [14], or primer and template independent ab initio DNA synthesis.

Among the diverse profiling applications, Kraken and MetaPhlAn are the most widely used in the field. Overall, while Kraken can achieve better overall performance, with higher precision and more accurate alpha- and beta-diversity measures, the trade-off between quality and required computational resources overall favors MetaPhlAn [15].

We propose here a quick method to carry out metagenomic reads profiling using MetaPhlan (v 4.1). Its online documentation (https://github.com/biobakery/MetaPhlAn/wiki/MetaPhlAn-4

) provides Conda installation and usage instructions, as well as tools for merging result output tables from different samples and plotting heatmaps and cladograms. First execution takes longer because it automatically downloads and sets up the microbial markers database (currently v. 31). We used the option —unclassified_estimation to quantify the reads not identified in the database, which could be assigned as amplification artifacts. The unmapped reads will be included in the SAM output file only if you add the option —no_map. Then, you can extract them from the BAM output file with flag -F 2 using Samtools (v. 1.21), as we suggest in the example below in order to extract the unmapped reads in a separate fasta file for further analysis.

```
#Example for the piPolB+D sample (pipolBD), consider a bash
script for multiple samples.
metaphlan pipolBD_R1_pair.fastq.gz,pipolBD_R2_pair.fastq.gz,
pipolBD_R1_unpair.fastq.gz,pipolBD_R2_unpair.fastq.gz --no_-
map --bowtie2out metaphlan_results/pipolBD_metaphlan.bowtie2.
bz2 --samout metaphlan_results/pipolBD_metaphlan4.sam --nproc
12 --input_type fastq -o metaphlan_results/pipolBD_profiled.
txt --unknown_estimation
#Extraction of the unmapped reads (if any) to a multifasta file
samtools view -u -F 2 metaphlan_results/pipolBD_metaphlan4.sam
| samtools fasta > metaphlan_results/pipolBD_metaphlan4_un-
mapped.fasta
```

Stacked barplots and heatmaps are the most useful plots to represent profiling data. As shown in Fig. 2, we could easily identify the similarities among our samples from different amplification methods. Furthermore, metagenomic profiling allowed us to identify a great ratio of unknown DNA in samples from piPolB-MDA and piPolB-MDA + D protocols, which we later argue may have originated from piPolB ab initio DNA synthesis (more details in [11]).

The detection of ab initio DNA synthesis also correlates with low complexity sequences, which can be analyzed using BBduk (from JGI BBtools) to quantify low-entropy reads, typically with a threshold of 0.5 bits.

Note

To remove ab initio sequences, we also used fastp (v.0.23.1) to filter out low complexity sequences (option -Y). Unfortunately, since those reads could not be de novo assembled, a detailed analysis of ab initio DNA replication by piPolB was not feasible from short reads.

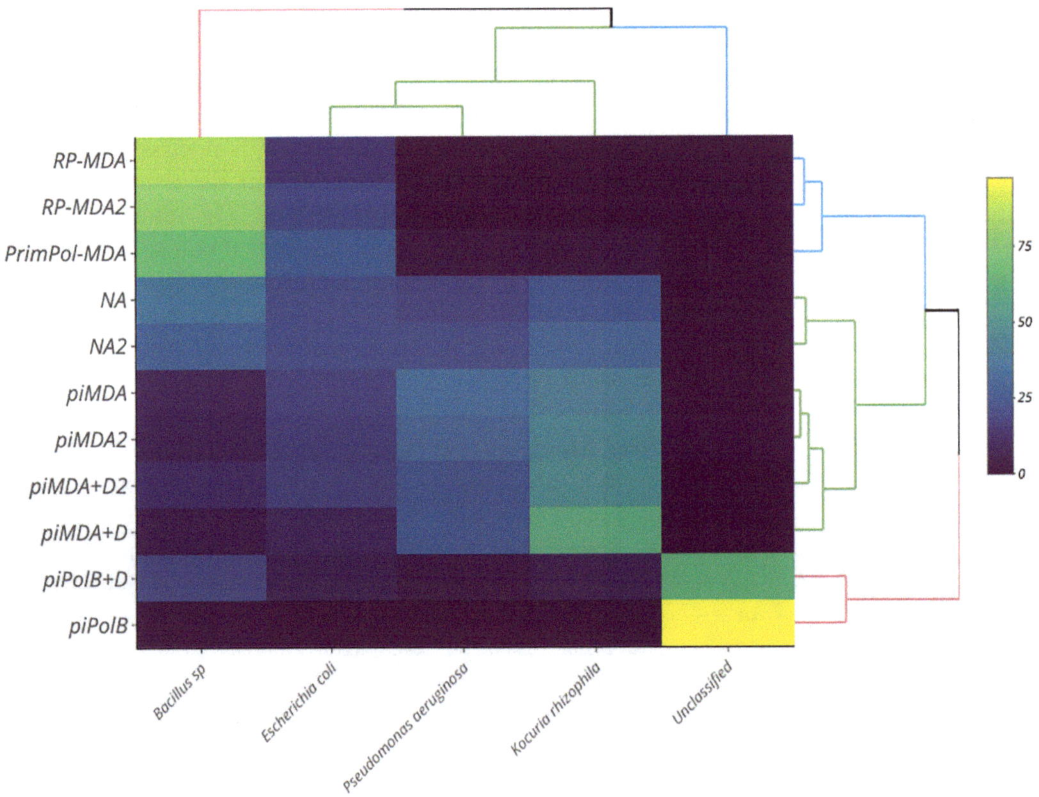

Fig. 2 Sample clustering from [11] by the MetaPhlAn profiling results. The R code of this and other alternative plots made with R (packages ggplot2 and heatmaply) are available in the GitHub repository

2.4 Step 4: Coverage Depth and Breadth

If a reference metagenome or genome is available for your input sample, mapping your reads against it will enable a comprehensive analysis of both sequencing and amplification bias. This type of analysis is particularly valuable when testing new amplification methods against well-established procedures, usually in the context of a previously characterized sample or, as demonstrated in [11], a mock community. Alternatively, control non-amplified samples can serve as a reference to assess MDA proficiency in a comparative manner.

Reads can be aligned with Bowtie2 [16] or another aligner to map the processed reads against a reference metagenome to obtain diverse parameters related to the coverage. Two pipelines, WeeSAM and Alfred [17], were tested. Both provide descriptive statistics of read alignment, but Alfred also includes mismatch and indel rates. Besides, WeeSAM provides a *Variation coefficient* that indicates variability in coverage (i.e., the evenness of coverage along the reference (meta)genome); a coefficient of variation less than 1 suggests low-variance coverage, which is considered favorable, while a coefficient greater than 1 would indicate high-variance of coverage values across the genome. Recent tools like BAMDash can also be used to obtain similar parameters and very convenient interactive visualizations.

Overall alignment rate of your sample against the reference (meta)genome can help identify samples with possible contaminants or unknown DNA, similar to profiling tools (Step 3). Moreover, coverage depth and breadth of each contig or genome in your reference sample may reveal amplification bias. For example, in [11], control non-amplified samples showed even coverage, whereas samples amplified with the traditional random-primers MDA (RP-MDA) protocol exhibited low breadth and depth coverage for genomes from *P. aeruginosa* and *K. rhizophila*, suggesting impaired amplification of high-GC sequences. Additionally, the *Variation coefficient* was higher in high-GC sequences in RP-MDA and notably in samples from piPolB-MDA, which contained a significant proportion of unknown DNA and over-amplification of low complexity sequences.

The following code illustrates the use of WeeSAM and Alfred after mapping sequencing reads against one or more reference genomes (4 in this case). The generated reports can be parsed with R to compute statistical metrics and plot individual parameters for each sample.

```
#Be careful to use the correct path in each file
#Reference genomes are in a directory named "refs"
# Step 1: build index for each reference genome
bowtie2-build refs/Kocuria_rhizophila_ATCC_9341.fasta kocuria
bowtie2-build refs/CP013113.1_PAER4.fasta PAER4
bowtie2-build refs/NC_000913.3.fasta coli
bowtie2-build refs/NC_000964.3.fasta sub
# Step 2: bowtie mapping (default is --end-to-end)
bowtie2 -x coli -1 pipolBD_R1_pair.fastq -2 pipolBD_R2_pair.
fastq -U pipolBD_R1_unpair.fastq,pipolBD_R2_unpair.fastq -p
12 | samtools view -bS - | samtools sort -@ 12 -o mapping_pi-
polBD_result_coli_sorted.bam
bowtie2 -x subtilis -1 pipolBD_R1_pair.fastq -2 pipolBD_R2_-
pair.fastq -U pipolBD_R1_unpair.fastq,pipolBD_R2_unpair.fastq
-p 12 | samtools view -bS - | samtools sort -@ 12 -o
mapping_pipolBD_result_subtilis_sorted.bam
bowtie2 -x PAER4 -1 pipolBD_R1_pair.fastq -2 pipolBD_R2_pair.
fastq -U pipolBD_R1_unpair.fastq,pipolBD_R2_unpair.fastq -p
12 | samtools view -bS - | samtools sort -@ 12 -o mapping_pi-
polBD_result_PAE_sorted.bam
bowtie2 -x kocuria -1 pipolBD_R1_pair.fastq -2 pipolBD_R2_-
pair.fastq -U pipolBD_R1_unpair.fastq,pipolBD_R2_unpair.fastq
-p 12 | samtools view -bS - | samtools sort -@ 12 -o
mapping_pipolBD_result_kocuria_sorted.bam
```

```
# Step 3: WeeSAM with html report (use --out for a flat table)
weeSAM --bam mapping_pipolBD_result_coli_sorted.bam --html
pipolBD_result_coli.html
weeSAM --bam mapping_pipolBD_result_subtilis_sorted.bam --
html pipolBD_result_subtilis.html
weeSAM --bam mapping_pipolBD_result_PAE_sorted.bam --html pi-
polBD_result_PAE.html
weeSAM --bam mapping_pipolBD_result_kocuria_sorted.bam --html
pipolBD_result_kocuria.html
# Step 3b: Alfred stats computing
alfred qc -r refs/NC_000913.3.fasta -o pipolBD_alfred_coli.
tsv.gz mapping_pipolBD_result_coli_sorted.bam
alfred qc -r refs/NC_000964.3.fasta -o pipolBD_alfred_subti-
lis.tsv.gz mapping_pipolBD_result_subtilis_sorted.bam
alfred qc -r refs/CP013113.1_PAER4.fasta -o pipolBD_alfred_-
PAE.tsv.gz mapping_pipolBD_result_PAE_sorted.bam
alfred qc -r refs/Kocuria_rhizophila_ATCC_9341.fasta -o pi-
polBD_alfred_kocuria.tsv.gz mapping_pipolBD_result_kocuria_-
sorted.bam
```

2.5 Step 5: GC Bias

Genomes contain sequences with varying ratios of GC/AT bases that impact DNA structure and stability, as well as the efficiency of amplification and sequencing methods. Similarly, metagenomes contain genomes that may exhibit a broad range of GC content, which often results in the underrepresentation of GC-rich and GC-poor species relative to those with average GC content. While sequencing technology influences GC bias, particularly during the library preparation step [18], the even representation of sequences with diverse GC content is challenging after WGA, especially for high-GC genomes [19]. Among the tools used to calculate GC bias, two commonly employed approaches exist. The first, implemented in the *computeGCBias* method [20] in deepTools [21], assumes a unimodal GC bias, meaning both GC-rich and GC-poor regions are underrepresented in read coverage. However, reports indicate that the pattern of GC bias can be diverse and not necessarily symmetric, depending on the sample nature and experimental procedures. To better capture diverse GC bias, a linear model was proposed to better capture the behavior of GC bias [22]. This alternative approach is implemented in the *CollectGcBiasMetrics* function in the Picard tool, developed by the Broad Institute (https://broadinstitute.github.io/picard/). Picard, a Java-based tool, can provide results in different formats, including a PDF report with plots and tables containing various metrics for further analysis. These metrics include the frequency of sequences at specific GC content defined by sequence length, typically 100 bp windows, and the normalized coverage by reads at particular GC content. Then, to quantify the bias specific to each MDA protocol, we performed pairwise comparisons between samples, involving control non-amplified samples (NA) and amplified samples. We considered only GC ranges with sufficient frequency in the reference sample (more than 5 windows). Sequences obtained using different MDA methods showed significant differences among themselves and compared to NA samples, except for the piMDA protocols. Correlation analysis of control samples and amplified samples revealed a more evenly distributed amplification in the piMDA method protocols (*see* Figure 6 in [11]). Picard is freely available for download from the Broad Institute's website and is thoroughly documented. GC bias calculation requires a prior BAM file of mapped reads against a reference (meta)genome, which can be generated using Bowtie2, as described earlier. The following example illustrates how to determine the GC bias of an amplified sample. Tables in plain text format can be merged and analyzed downstream.

```
#Example with piPolB+D sample
java -jar picard.jar CollectGcBiasMetrics \
#input BAM file
 I=bowtiemapping_pipolBD_metagenome_sorted.bam \
#detailed output
 O=gc/pipolBD_gc_bias_metagenome.txt \
#PDF report
 CHART=gc/pipolBD_gc_bias_metagenome.pdf \
#summary table
 S=gc/pipolBD_gc_bias_metagenome_summary_metrics.txt \
#reference genome
 R=refs/metagenome.fasta
```

2.6 Step 6: Metagenome Assembly

We recommend a streamlined approach for genome or metagenome de novo assembly utilizing SPAdes (v. >3.16) [23], followed by qualitative and quantitative evaluation with Quast (v. >5.1.0rc1) [24]. Our tests using SPAdes in normal mode, single-cell mode (--sc), and meta mode (--meta) indicated that the meta mode produced the most optimal results across all our samples, as evidenced by N50 and L50 metrics.

Assessment of assemblies using Quast and MetaQuast not only enables a quantitative comparison against reference genomes but also provides a comprehensive report on read coverage, mismatches, assembly k-mer complexity (option -k), and gene annotation statistics (-f and --mgm or --glimmer), among other possibilities. Additionally, it includes interactive plots and circular comparative Circos plots (--circos). While multiple samples can be processed in a single command, performing a full comparative analysis that includes reference genomes, raw reads, and the assembly is computationally intensive. Therefore, either opting for a comparative assessment without the reads or individual sample analysis might be advisable.

```
#Asssembly with spades
spades.py --pe1-1 pipolBD_R1_pair.fastq.gz --pe1-2 pi-
polBD_R2_pair.fastq.gz --pe1-s pipolBD_R1_unpair.fastq.gz --
pe1-s pipolBD_R2_unpair.fastq.gz -t 32 --meta -o spades_re-
sults/F2
#Quast comparative assessment of all samples including raw
reads
#Using '-l' you can customize the labels of your samples
python3 metaquast.py \
-r refs/NC_000964.3.fna,refs/GCF_000005845.2_ASM584v2_ge-
nomic.fna,refs/Kocuria_rhizophila_ATCC_9341.fasta,refs/
GCF_000006765.1_ASM676v1_genomic.fna -k -f --mgm --circos -t
32 \
-l "NA, NA2, RP-MDA, RP-MDA2, PrimPol-MDA, piPolB+D, piMDA,
piMDA2, piMDA+D, piMDA+D2" \
spades_results/Ctrl/contigs.fasta --pe1 trimmed/Ctrl_R1_pair.
fastq.gz --pe2 trimmed/Ctrl_R2_pair.fastq.gz -o Ctrl \
spades_results/Ctrl2/contigs.fasta --pe1 trimmed/
Ctrl2_R1_pair.fastq.gz --pe2 trimmed/Ctrl2_R2_pair.fastq.gz
-o Ctrl2 \
spades_results/N4/contigs.fasta --pe1 trimmed/RPMDA_R1_pair.
fastq.gz --pe2 trimmed/RPMDA_R2_pair.fastq.gz -o RPMDA \
spades_results/RPMDA/contigs.fasta --pe1 trimmed/RPMDA2_R1_-
pair.fastq.gz --pe2 trimmed/RPMDA2_R2_pair.fastq.gz -o RPMDA2
\
spades_results/ppolmda/contigs.fasta --pe1 trimmed/ppolm-
da_R1_pair.fastq.gz --pe2 trimmed/ppolmda_R2_pair.fastq.gz -o
PrimpolMDA \
spades_results/pipolBD/contigs.fasta --pe1 trimmed/pi-
polBD_R1_pair.fastq.gz --pe2 trimmed/pipolBD_R2_pair.fastq.gz
-o F2 \
spades_results/pimda/contigs.fasta --pe1 trimmed/pimda_R1_-
pair.fastq.gz --pe2 trimmed/pimda_R2_pair.fastq.gz -o piMDA \
spades_results/pimda2/contigs.fasta --pe1 trimmed/pimda2_R1_-
pair.fastq.gz --pe2 trimmed/pimda2_R2_pair.fastq.gz -o piMDA2
\
spades_results/pimdaD/contigs.fasta --pe1 trimmed/pimdaD_R1_-
pair.fastq.gz --pe2 trimmed/pimdaD_R2_pair.fastq.gz -o piMDAD
\
spades_results/pimdaD2/contigs.fasta --pe1 trimmed/pim-
daD2_R1_pair.fastq.gz --pe2 trimmed/pimdaD2_R2_pair.fastq.gz
-o piMDAD2
```

2.7 Step 7: Metagenome Annotation

In terms of comparison of WGA methods, the number of annotated features reflects a high quality of the assembly. This can be obtained directly from Quast. Additionally, we propose a more detailed annotation of the metagenome with Bakta [25], which annotates CDS, RNA genes, small and non-coding RNAs, and CRISPR arrays, as well as short ORFs and pseudogenes. Moreover, for those interested in gene functions, Batka provides a large and frequently updated database.

> **Tip**
>
> Bakta has various dependencies and third-party tools that need to be installed. Therefore, it is highly recommended to create a dedicated Conda environment. Alternatively, a maintained Docker image is also available. For detailed installation instructions, refer to the Bakta repository. The Bakta database (currently version 6, released on February 25, 2025) is available in two versions: Light (3.9 GB) and full (84 GB). It is up to the user which one suits their needs better.

2.8 Step 8: Variant Calling

Variant calling with Snippy (v. 4.6.0) is a straightforward and precise method to detect genetic variants in a genome or metagenome assembly [26]. In the context of a mock metagenome, the emergence of novel variants absent from the control sequences can offer valuable insights into the accumulation of amplification errors.

SNP identification requires a reference genome, either in FASTA or GBK format, along with paired-end sequencing reads. Similarly, one can provide the assembled metagenome and the trimmed paired reads. To facilitate comparison, instead of the reference metagenome, we used the metagenome assembly obtained with SPAdes and the control samples as a reference.

```
mkdir snippy_results
#Variants calling example in NA and piMDA samples
snippy --cpus 24 --outdir snippy_results/ctrl_snp --ref spa-
des_results/Ctrl/contigs.fasta --R1 trimmed/Ctrl_R1_paired.
fastq.gz --R2 trimmed/Ctrl_R2_paired.fastq.gz
snippy --cpus 24 --outdir snippy_results/pimda --ref spades_-
results/Ctrl/contigs.fasta --R1 trimmed/pimda_R1_paired.
fastq.gz --R2 trimmed/pimda_R2_paired.fastq.gz
#Creation of a comparative table between control and piMDA
grep -v -wf Ctrl_snp/snps.csv pimda_snp/snps.csv > NA_pimda_-
comparison.csv
```

The output includes various files suitable for downstream analysis, such as summary tables and variant data in VCF, GFF, and BED formats, along with the alignments. These tables can be easily parsed and, as shown in the example below, utilized to compare and identify missing and newly identified variants in the amplified samples.

2.9 Step 9: Sequence Complexity and K-Mer Distribution

Although genomes contain regions of low complexity sequences and repeats, DNA amplification and sequencing artifacts can result in bias or an increase of those patterns. In fact, as discussed above, unknown, ab initio DNA that can arise from ab initio DNA synthesis by diverse DNA polymerases [11, 27, 28] is usually enriched in low complexity sequences. Trimming of raw Illumina reads with Trimmomatic or fastp is usually sufficient to remove homopolymeric and other low complexity sequences. Additionally, Joint Genome Institute's [BBtools] (https://github.com/BioInfoTools/BBMap/), named BBduk, allows accurate quantification of low complexity sequences in a fasta or fastq file using a custom entropy threshold. We used it to analyze the MetaPhlAn unmapped reads, finding a high ratio of low complexity reads in piPolB-MDA samples, consistent with ab initio DNA amplification. Additionally, to study the mechanism of untemplated ab initio DNA synthesis, the abundance of short di-, tri-, or tetranucleotides can be very informative [27].

The distribution pattern of k-mers can be used to compare sequences in various ways. Long k-mers can be very specific and enable accurate metagenome profiling [29] or binning [30]. Additionally, the abundance of k-mers can indicate biased overamplification. Comparison of k-mer spectra can provide useful information in this context. The K-mer Analysis Toolkit (KAT) allows multiple comparisons between samples, GC content, and also between each sample and a reference genomic sequence [31]. Regarding the effect of DNA amplification, the assembly spectra copy number plot produced by the comp tool provides information on the extent and type of k-mers from the reads that are included in the assembly. It decomposes the k-mer spectrum of a read dataset based on the frequency at which the k-mers are encountered in the assembly. This method represents how many elements of each frequency in the read spectrum ended up included only once in the assembly, how many were included twice, etc. Additionally, it records how many k-mers do not appear in the assembly. If an assembler performed appropriately on your reads, the ratio of reads-specific k-mers (BLACK in the output plots as in Fig. 3) would be low, while genuine unique content appears exactly once (RED, for haploid genomes) and repeated content at higher levels of duplication presents as distributions centered around multiples of the sampling frequency for unique content.

```
#Example for NA, Random-primers MDA and piMDA+d samples
kat comp -t 24 -o kat_test -m16 'trimmed/Ctrl_R1_paired.fastq.gz trimmed/Ctrl_R2_paired.fastq.gz' spades_results/Ctrl/contigs.fasta
kat comp -t 24 -o kat_test -m16 'trimmed/RPMDA_R1_paired.fastq.gz trimmed/RPMDA_R2_paired.fastq.gz' spades_results/RPMDA/contigs.fasta
```

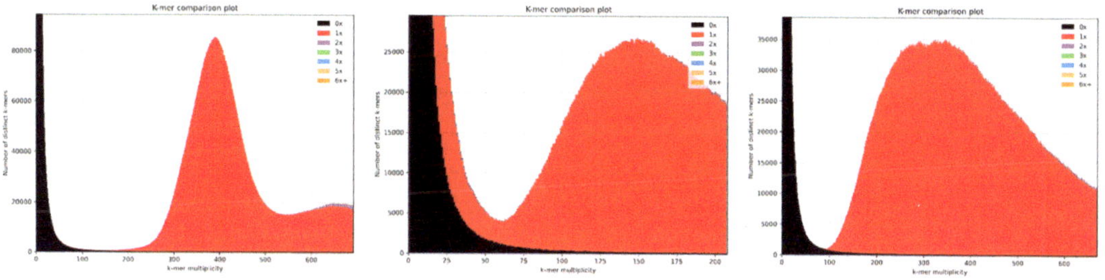

Fig. 3 KAT comparative spectra can be utilized to verify the coherence of assemblies relative to the content within the reads that were used to generate each assembly. The colors indicate the number of times k-mer content is present in the assembly, as specified in the legend. Despite the different axis scales, there are noticeably more k-mers present exclusively in the reads (black) of the amplified samples (center panel and rightmost), compared to the control sample (left panel). For detailed results, please refer to Figure S8 in [11]

```
kat comp -t 24 -o kat_test -m16 'trimmed/pimdaD_R1_paired.
fastq.gz trimmed/pimdaD_R2_paired.fastq.gz' spades_results/
pimdaD/contigs.fasta
```

2.10 Step 10: Metagenome Binning

Binning, i.e., reference-free grouping of contigs from assemblies facilitates evaluation of whether amplification has impacted the extraction of metagenome-assembled genomes. The MetaCoAG tool can be employed to generate contig bins incorporating abundance, base composition, and connectivity information from assembly graphs, thereby producing high-quality bins [32], which would be subsequently assigned to their corresponding reference genome.

MetaCoAG (v. 1.2.2) supports contigs assembled using metaSPAdes and MEGAHIT. The assembly graph and a table containing contig average coverage must also be provided. This table can be easily obtained with CoverM. Distances between all the computed bins and the reference genomes can be quickly computed using Mash [33]. This complements the previous analysis of the metagenome assembly to identify which genomes in the metagenome were poorly assembled due to amplification bias and errors.

```
#MetaCoAG and CoverM from Conda
#Step 1: Obtain coverage per contig table
coverm contig -1 trimmed/Ctrl_R1_paired.fastq.gz -2 trimmed/
Ctrl_R2_paired.fastq.gz -r spades_results/Ctrl/contigs.fasta
-o binning/abundance_Ctrl.tsv -t 12 -v
#Step 2: MetaCoAG bining
MetaCoAG --assembler spades --graph spades_results/Ctrl/assem-
bly_graph_with_scaffolds.gfa --contigs spades_results/Ctrl/
contigs.fasta --paths spades_results/Ctrl/contigs.paths --
abundance binning/abundance_Ctrl.tsv --output binning/Ctrl –
mg_threshold 0.4 –bin_mg_threshold 0.2 --nthreads 12
```

```
#Step 3: Match bins with references
#3.1. Create references sketch
mash sketch -o reference refs/NC_000913.3.fasta refs/
NC_000964.3.fasta refs/CP013113.1_PAER4.fasta refs/Kocuria_r-
hizophila_ATCC_9341.fasta
#3.2. Create sample sketch for all the bins
mash sketch -o ctrl binning/Ctrl/bins/bin_0.fasta binning/
Ctrl/bins/bin_1.fasta binning/Ctrl/bins/bin_2.fasta binning/
Ctrl/bins/bin_3.fasta binning/Ctrl/bins/bin_4.fasta binning/
Ctrl/bins/bin_5.fasta binning/Ctrl/bins/bin_6.fasta
#3.3. Compute distances
mash dist reference.msh ctrl.msh > ctrl_dist.txt
```

Then, we plotted the size of the bins in stacked bars colored by the best reference. For our mock metagenome, distance <0.1, $p < 10^{-10}$, and length > 10 were enough to assign references.

Funding

Our lab is funded by MCIN/AEI/10.13039/501100011033 and *ERDF A way of making Europe* [PID2021-123403NB-I00]. E.D.L.E., C.M.C., and V.M.C. received PhD Fellowships from the Spanish Ministry of University [FPU23/01361], Ministry of Science and Innovation [PRE2019-087304], and UAM [SFPI/2023–00603], respectively.

References

1. van der Meij KRM, Sistermans EA, Macville MVE et al (2019) TRIDENT-2: National implementation of genome-wide non-invasive prenatal testing as a first-tier screening test in The Netherlands. Am J Hum Genet 105: 1091–1101. https://doi.org/10.1016/j.ajhg.2019.10.005
2. Yamaguchi I, Watanabe T, Ohara O, Hasegawa Y (2019) PCR-free whole exome sequencing: cost-effective and efficient in detecting rare mutations. PLoS One 14:e0222562. https://doi.org/10.1371/journal.pone.0222562
3. Shi Y, Wang G, Lau HC-H, Yu J (2022) Metagenomic sequencing for microbial DNA in human samples: emerging technological advances. Int J Mol Sci 23:2181. https://doi.org/10.3390/ijms23042181
4. Costea PI, Zeller G, Sunagawa S et al (2017) Towards standards for human fecal sample processing in metagenomic studies. Nat Biotechnol 35:1069–1076. https://doi.org/10.1038/nbt.3960
5. Knudsen BE, Bergmark L, Munk P et al (2016) Impact of sample type and DNA isolation procedure on genomic inference of microbiome composition. mSystems 1:e00095–16. https://doi.org/10.1128/mSystems.00095-16
6. Poulsen CS, Ekstrøm CT, Aarestrup FM, Pamp SJ (2022) Library preparation and sequencing platform introduce bias in metagenomic-based characterizations of microbiomes. Microbiol Spectr
7. Ordóñez CD, Redrejo-Rodríguez M (2023) DNA polymerases for whole genome amplification: considerations and future directions. Int J Mol Sci 24:9331. https://doi.org/10.3390/ijms24119331
8. Povilaitis T, Alzbutas G, Sukackaite R et al (2016) In vitro evolution of phi29 DNA polymerase using isothermal compartmentalized self replication technique. Protein Eng Des Sel 29:617–628. https://doi.org/10.1093/protein/gzw052

9. Salas Falgueras M, Lázaro Bolos JM, De Vega José M, et al (2015) Bacteriophage phi29 dna polymerase variants having improved thermoactivity
10. Picher AJ, Budeus B, Wafzig O et al (2016) TruePrime is a novel method for whole-genome amplification from single cells based on TthPrimPol. Nat Commun 7:13296. https://doi.org/10.1038/ncomms13296
11. Ordóñez CD, Mayoral-Campos C, Egas C, Redrejo-Rodríguez M (2023) A primer-independent DNA polymerase-based method for competent whole-genome amplification of intermediate to high GC sequences. NAR Genom Bioinform 5:lqad073. https://doi.org/10.1093/nargab/lqad073
12. Tan Y, Wang R, Wu B et al Retrieval-enhanced mutation mastery: Augmenting zero-shot prediction of protein language model. https://doi.org/10.48550/arXiv.2410.21127
13. Redrejo-Rodríguez M, Ordóñez CD, Berjón-Otero M et al (2017) Primer-independent DNA synthesis by a family b DNA polymerase from self-replicating mobile genetic elements. Cell Rep 21:1574–1587. https://doi.org/10.1016/j.celrep.2017.10.039
14. Alsmadi O, Alkayal F, Monies D, Meyer BF (2009) Specific and complete human genome amplification with improved yield achieved by phi29 DNA polymerase and a novel primer at elevated temperature. BMC Res Notes 2:48. https://doi.org/10.1186/1756-0500-2-48
15. Wright RJ, Comeau AM, Langille MGI (2023) From defaults to databases: parameter and database choice dramatically impact the performance of metagenomic taxonomic classification tools. Microb Genom 9:000949. https://doi.org/10.1099/mgen.0.000949
16. Langmead B, Wilks C, Antonescu V, Charles R (2019) Scaling read aligners to hundreds of threads on general-purpose processors. Bioinformatics 35:421–432. https://doi.org/10.1093/bioinformatics/bty648
17. Rausch T, Hsi-Yang Fritz M, Korbel JO, Benes V (2019) Alfred: interactive multi-sample BAM alignment statistics, feature counting and feature annotation for long- and short-read sequencing. Bioinformatics 35:2489–2491. https://doi.org/10.1093/bioinformatics/bty1007
18. Sato MP, Ogura Y, Nakamura K et al (2019) Comparison of the sequencing bias of currently available library preparation kits for Illumina sequencing of bacterial genomes and metagenomes. DNA Res 26:391–398. https://doi.org/10.1093/dnares/dsz017
19. Pinard R, Winter A de, Sarkis GJ, et al (2006) Assessment of whole genome amplification-induced bias through high-throughput, massively parallel whole genome sequencing. BMC Genomics 7:216. https://doi.org/10.1186/1471-2164-7-216
20. Benjamini Y, Speed TP (2012) Summarizing and correcting the GC content bias in high-throughput sequencing. Nucleic Acids Res 40:e72. https://doi.org/10.1093/nar/gks001
21. Ramírez F, Dündar F, Diehl S et al (2014) deepTools: a flexible platform for exploring deep-sequencing data. Nucleic Acids Res 42: W187–W191. https://doi.org/10.1093/nar/gku365
22. Chen Y-C, Liu T, Yu C-H et al (2013) Effects of GC bias in next-generation-sequencing data on de novo genome assembly. PLoS One 8: e62856. https://doi.org/10.1371/journal.pone.0062856
23. Nurk S, Meleshko D, Korobeynikov A, Pevzner PA (2017) metaSPAdes: a new versatile metagenomic assembler. Genome Res 27: 824–834. https://doi.org/10.1101/gr.213959.116
24. Mikheenko A, Saveliev V, Gurevich A (2016) MetaQUAST: evaluation of metagenome assemblies. Bioinformatics 32:1088–1090. https://doi.org/10.1093/bioinformatics/btv697
25. Schwengers O, Jelonek L, Dieckmann MA et al (2021) Bakta: rapid and standardized annotation of bacterial genomes via alignment-free sequence identification. Microb Genom 7: 000685. https://doi.org/10.1099/mgen.0.000685
26. Bush SJ, Foster D, Eyre DW et al (2020) Genomic diversity affects the accuracy of bacterial single-nucleotide polymorphism-calling pipelines. GigaScience 9:giaa007. https://doi.org/10.1093/gigascience/giaa007
27. Castle SD, Woolfson A, Linshiz G et al Analysis and control of untemplated DNA polymerase activity for guided synthesis of kilobase-scale DNA sequences. https://doi.org/10.1101/2024.08.29.610422
28. Zyrina NV, Antipova VN, Zheleznaya LA (2014) Ab initio synthesis by DNA polymerases. FEMS Microbiol Lett 351:1–6. https://doi.org/10.1111/1574-6968.12326
29. Liu S, Koslicki D (2022) CMash: fast, multi-resolution estimation of k-mer-based Jaccard and containment indices. Bioinformatics 38: i28–i35. https://doi.org/10.1093/bioinformatics/btac237

30. Kyrgyzov O, Prost V, Gazut S et al (2020) Binning unassembled short reads based on k-mer abundance covariance using sparse coding. Gigascience 9:giaa028. https://doi.org/10.1093/gigascience/giaa028
31. Mapleson D, Garcia Accinelli G, Kettleborough G et al (2017) KAT: a K-mer analysis toolkit to quality control NGS datasets and genome assemblies. Bioinformatics 33:574–576. https://doi.org/10.1093/bioinformatics/btw663
32. Mallawaarachchi V, Lin Y (2022) Accurate binning of metagenomic Contigs using composition, coverage, and assembly graphs. J Comput Biol 29:1357–1376. https://doi.org/10.1089/cmb.2022.0262
33. Ondov BD, Treangen TJ, Melsted P et al (2016) Mash: fast genome and metagenome distance estimation using MinHash. Genome Biol 17:132. https://doi.org/10.1186/s13059-016-0997-x

Chapter 11

Sequencing RSV Whole Genome Using a Long Amplicon-Based Method with Oxford Nanopore Technologies

Xiaomin Dong, Steven Edwards, Yi-Mo Deng, Clyde Dapat, Arada Hirankitti, and Ian G. Barr

Abstract

Respiratory syncytial virus (RSV) infection continues to be a significant burden on public health care systems and is a global health concern. Whole genome sequencing (WGS) provides a useful tool to better understand the viral transmission and emerging mutations that may impact antibody treatments, antiviral drug sensitivity, and vaccine effectiveness. Here, we describe a rapid and sensitive protocol for sequencing clinical samples of both human RSV-A and RSV-B viruses based on the Oxford Nanopore Technology (ONT) sequencing platform. It involves long amplicon generation by setting up two one-step multiplex reverse-transcription polymerase chain reactions (mRT-PCR) for each sample, library preparation with the ONT rapid barcoding kit and NGS data analysis with the ARTIC pipeline.

Key words Respiratory syncytial virus, Whole genome sequencing, ONT, Rapid barcoding

1 Introduction

Respiratory syncytial virus (RSV) is a negative-sense, single-stranded, non-segmented RNA virus of approximately fifteen thousand two hundred bases, belonging to the *Pneumoviridae* family [1]. There is a single RSV serotype with two major antigenic subgroups or subtypes, RSV A and RSV B, which share around 46.7–96.4% homology at the amino acid level [2]. Strains of both subtypes often co-circulate, but generally, one of the subtypes predominates each season and can vary from country to country. Sequencing the entire RSV genome is important for monitoring viral evolution and the rise and spread of mutations that might generate resistance to preventatives and therapeutics against RSV [3]. We developed a simplified amplicon-based one-step multiplex reverse-transcription polymerase chain reaction (mRT-PCR) and

Yafeng Ma (ed.), *Whole Genome Amplification: Methods and Protocols*, Methods in Molecular Biology, vol. 3003, https://doi.org/10.1007/978-1-0716-5060-8_11,

combined it with the rapid and sensitive Oxford Nanopore Technologies (ONT) NGS workflow for WGS of both human RSV-A and RSV-B viruses [4]. This method utilizes one-step RT-PCR reagents to amplify six overlapping amplicons to have the RSV whole genome fully covered in two separate RT-PCR reactions. To improve the efficiency of RSV viral genome amplification and sequencing, DNase treatment of the extracted RNA is recommended for the removal of host DNA genomes in clinical samples. Additionally, the use of Oxford Nanopore Technology (ONT) and a rapid barcoding library preparation kit enables a simple, rapid, and cost-effective solution for sequencing RSV genomes [5]. For NGS data analysis, the ARTIC pipeline is recommended for better assembly of ONT reads and removal of primer sequences, as it was originally designed for analyzing viral nanopore sequencing data based on PCR-tiling amplicon methods.

2 Materials

2.1 Clinical Specimens

RSV-positive respiratory samples including nasal swabs, nasopharyngeal swabs, nasal washes, or nasopharyngeal aspirates can be used for the described RSV whole genome sequencing method. Samples are supposed to be stored at −80 °C until processing.

2.2 Viral RNA Extractions

For viral RNA extraction from clinical samples, QIAamp Viral RNA Kits (QIAGEN) can be used for manual extractions, or QIAamp 96 Virus QIAcube HT (QIAGEN) can be used for automated extractions.

2.3 DNase Treatment

1. ezDNase (Invitrogen).
2. 13.2 mM EDTA.
3. Thin-walled PCR 0.2 mL strip tube or 96-well thermal cycling plate.
4. A thermal cycler.
5. Microcentrifuge.

2.4 One-Step Multiplex Reverse Transcription Polymerase Chain Reaction (RT-PCR) for RSV WGS

1. The SuperScript IV One-Step RT-PCR System (Invitrogen).
2. A thermal cycler.
3. Thin-walled PCR 0.2 mL strip tube or 96-well thermal cycling plate.
4. DNA oligonucleotides used as primers in Table 1.
5. Microcentrifuge.
6. Reconstitute a 100 μM stock solution for each primer that has been dried into a powder form with nuclease-free water. Find

Table 1
Primers used in the one-step multiplex RT-PCR

Primer pools	Primers	Sequence (5′-3′)
1	1-1F	ACGCGAAAAAATGCGTACTACAAAC
	1-1R	CTGMACCATAGGCATTCATAAACA
	1-3F3	GCATCACTWACAATATGGGTKCC
	1-3R6	CACTTTTGATYTTGTTCACTTCYCC
	1-5F	TGATGCATCAATATCTCAAGTCA
	1-5R	GRCCTATDCCTGCATACTC
2	2-2F	ATGGGAGARGTRGCTCCAGAATA
	2-2R	CGTGTAGCTGTRTGYTTCCAA
	2-4F	AGCAAATTYTGGCCYTAYTTTAC
	2-4R1	CTCATAGCAACACATGCTGATTG
	2-4R2	GAGTTTGCTCATGGCAACACAT
	2-6F	TGGACCATWGAAGCYATATCA
	2-6R	AGTGTCAAAAACTAATRTCTCGT
	2-6R2	ACGAGAAAAAAAGTGTCAAAAACTAAT

Table 2
The volume of each primer needed to prepare one reaction of primer pool 1

Primer pool 1	Volumes (μL) of primer working solution for each reaction
1-1F	0.2
1-1R	0.2
1-3F3	0.69
1-3R6	0.69
1-5F	0.75
1-5R	0.75

the amount of DNA oligonucleotides in nmol on the tube label. Multiplying this number by 10 is the volume of nuclease-free water in μL needed to prepare a 100 μM stock solution. The resulting product should be stored at −20 °C and is stable for 2 years. Dilute one volume of storage solution with 4 volumes of nuclease-free water to get a 20 μM of working solution for each primer.

7. Prepare primer pool 1 and pool 2 in a 1.5-mL microcentrifuge tube. Use the volumes listed in Tables 2 and 3 below. Multiply each volume by the number of samples being processed with a 10% overfill.

Table 3
The volume of each primer needed to prepare one reaction of primer pool 2

Primer pool 2	Volumes (μL) of primer working solution for each reaction
2-2F	0.31
2-2R	0.31
2-4F	0.75
2-4R1	0.38
2-4R2	0.38
2-6F	0.55
2-6R	0.55
2-6R2	0.55

2.5 Visualization of RT-PCR Products

1. Agilent 4200 TapeStation instrument (Agilent).
2. D5000 ScreenTape (Agilent).
3. D5000 Reagents (Agilent).
4. Optical Tube 8x Strip and Optical Cap 8x Strip (Agilent).
5. Vortex mixer.
6. Microcentrifuge.

2.6 Quantification of RT-PCR Products with Qubit Fluorometer

1. Qubit 1X dsDNA HS Assay Kits (Thermo Fisher).
2. Qubit fluorometer (Thermo Fisher).
3. 0.5 mL PCR tubes for the Qubit.
4. Vortex mixer.
5. Microcentrifuge.

2.7 ONT NGS

1. Rapid Barcoding Kit 96 V14 (Oxford Nanopore Technologies).
2. MinION Flow Cell R10 (Oxford Nanopore Technologies).
3. MinION sequencer (Oxford Nanopore Technologies).
4. A computer with a minimum of 4 cores/8 threads, 8 GB of GPU, and 16GB of RAM.
5. A thermal cycler.
6. A magnetic rack for 1.5 mL tubes or A magnetic plate for the 96-well thermal cycling plate.
7. 80% ethanol (freshly prepared).
8. Centrifuge.

2.8 ONT Data Analysis with ARTIC Pipeline

1. Version 1.5.1 ARTIC pipeline (a bioinformatics pipeline for working with virus sequencing data sequenced with nanopore) (https://github.com/artic-network/fieldbioinformatics).
2. RSV genome references and associated primer sequences bed files (https://github.com/aimer288/RSV-reference-sequences-for-WGS).
3. A computer with a minimum of 2 CPUs and 4GB of RAM.

3 Methods

3.1 Viral RNA Extraction from the Samples of Interest

Viral RNA is extracted from RSV-positive respiratory samples with QIAamp Viral RNA Kits for manual extractions or QIAamp 96 Virus QIAcube HT for automated extractions, by following the manufacturer's instructions (*see* **Note 1**).

3.2 DNase Treatment

DNase treatment to the viral RNA extracted above is recommended prior to reverse transcription and amplicon generation (*see* **Note 2**).

1. Thaw the 10X ezDNase Buffer before use, mix by flicking the tube, briefly centrifuge, and place on ice.
2. Briefly centrifuge and keep the ezDNase Enzyme on ice.
3. Add 17 μL of extracted RNA to a sterile thin-walled PCR 0.2 mL strip tube or well of a 96-well thermal cycling plate.
4. Add 2 μL of 10X ezDNase Buffer and 1 μL of ezDNase Enzyme to each RNA sample with a pipette tip changed between each addition (*see* **Note 3**).
5. Mix gently, centrifuge briefly, and incubate at 37 °C for 10 min (*see* **Note 4**).
6. Terminate the reaction by adding 4 μL of 13.2 mM EDTA, mixing gently, centrifuging briefly, and heating at 65 °C for 10 min.
7. After incubation, immediately place on ice to bring the sample to room temperature, then briefly centrifuge, and place the tube(s) or plate on ice.
8. Proceed to the next step.

3.3 RSV Amplicon Generation by One-Step mRT-PCR for RSV WGS

Each RSV sample requires two primer pools to generate six PCR fragments in two RT PCR reactions. There are three fragments produced in each RT-PCR reaction.

1. Thaw all required reagents on ice, then flick or invert the tube (s) to mix, and then pulse-spin.
2. Combine the following components in Tables 4 and 5 to prepare 25 μL of each multiplex RT-PCR reaction 1 and 2 in

Table 4
Components for making one reaction of multiplex RT-PCR reaction 1

Multiplex RT-PCR reaction 1	Volumes (μL) of one reaction
Primer pool 1	3.28
DNase-treated RNA	8.97
2X Platinum SuperFi RT-PCR Master Mix	12.5
SuperScript IV RT Mix	0.25
Total	25

Table 5
Components for making one reaction of multiplex RT-PCR reaction 2

Multiplex RT-PCR reaction 2	Volumes (μL) of one reaction
Primer pool 2	3.78
DNase-treated RNA	8.47
2X Platinum SuperFi RT-PCR Master Mix	12.5
SuperScript IV RT Mix	0.25
Total	25

a new thin-walled PCR 0.2 mL strip-tube or well of a new 96-well thermal cycling plate. Multiply each volume by the number of samples being processed with a 10% overfill.

3. Mix by flicking the tube(s) or inverting the plate until homogenized, then centrifuge briefly (*see* **Note 5**).
4. Thermocycling conditions are 10 min at 50 °C for reverse transcription, 2 min at 98 °C for reverse transcriptase inactivation, followed by 40 cycles of 98 °C for 10 s, 55°C for 30 s, and 72 °C for 3 min, a final extension step at 72 °C for 5 min, and a 4 °C hold. Set the temperature of the heated lid to 105 °C (*see* **Note 6**).

3.4 Visualization of RT-PCR Products

Three amplicons in each of both RT-PCR reactions are distinguishable by size, so visualization of RT-PCR products on TapeStation System can be used to confirm the successful fragment amplification (*see* **Note 7**).

1. Bring reagents to room temperature, and vortex before use, then centrifuge briefly.
2. Mix 10 μL of D5000 Sample Buffer with 1 μL of D5000 Ladder to prepare the DNA ladder solution in the first tube of an Agilent strip tube.

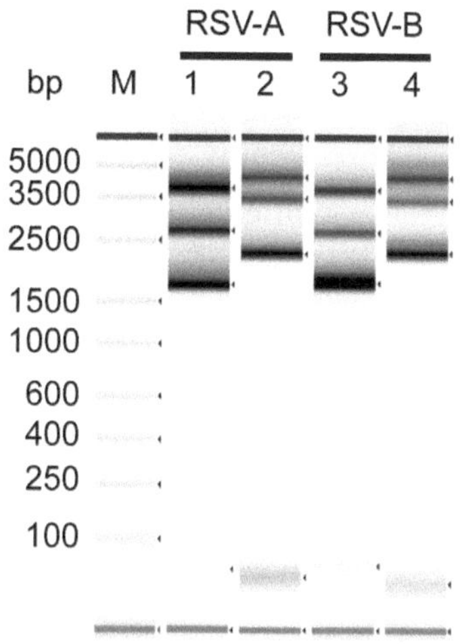

Fig. 1 A representative TapeStation image of two-pool PCR products of viral RNA from one RSV-A and one RSV-B positive clinical samples. Molecular marker (M); multiplex RT-PCR products of pool 1 (RSV-A in lane 1 and RSV-B in lane 3) contain ~1898 bp (fragment 1), ~2876 (fragment 3), and ~ 3936 bp (fragment 5) fragments; multiplex RT-PCR products of pool 2 (RSV-A in lane 2 and RSV-B in lane 4) contain ~2443 bp (fragment 2), ~3634 bp (fragment 4), and ~ 4354 bp (fragment 6) fragments.

3. Add 1 μL of each multiplex PCR reaction product to 10 μL of D5000 Sample Buffer to get a sample solution in the strip tube or 96-well plate.
4. Vortex both the DNA ladder and sample solutions for 1 min, then spin down to collect them at the bottom of the tube or plate (*see* **Note 8**).
5. Put samples on the 4200 TapeStation instrument and select sample positions with the 4200 TapeStation Controller Software for measurement. An image from the TapeStation is shown in Fig. 1.

3.5 Quantification of DNA Fragments for WGS

To obtain an even distribution of RT-PCR products as input for NGS library preparation, PCR products from two multiplex PCR reactions of each sample must be quantified and pooled in equal molarity. It is an easy way to measure the concentration of RT-PCR products with the Qubit fluorometer (*see* **Note 9**).

1. Bring reagents to room temperature, and vortex before use, then centrifuge briefly.
2. In two separate 0.5 mL thin-walled and clear PCR tubes, mix 10 μL of each Qubit standard with 190 μL of Qubit 1X dsDNA working solution.

3. Transfer 1 to 20 μL of each RT-PCR reaction product to a 0.5 mL thin-walled and clear PCR tube and make the final volume to 200 μL with Qubit 1X dsDNA working solution.
4. Vortex each tube vigorously for 3–5 s, spin down the tubes briefly, and incubate all tubes at room temperature for 2 min in the dark.
5. Proceed to read standards and samples.

3.6 Amplicon Normalizations for WGS

ONT recommends 50–200 ng of amplicons from every sample as input for library preparation when using the Rapid Barcoding Kit 96 V14 (SQK-RBK114.96) reagent.

1. If processing less than 20 samples, transfer 88 ng of RT-PCR reaction one products and 112 ng of RT-PCR reaction two products generated from each sample into a 0.2 mL PCR strip tube or an Eppendorf 96 LoBind PCR plate (*see* **Note 10**).
2. For more than 20 samples, transfer 22 ng of RT-PCR reaction one product and 28 ng of RT-PCR reaction two products generated from each sample into a 0.2 mL PCR strip tube or an Eppendorf 96 LoBind PCR plate.
3. Normalize RT-PCR products to make the final volume of each sample 10 μL with nuclease-free water (*see* **Note 11**).
4. Briefly vortex and spin down the tubes or plate briefly in a microfuge, then place on ice.

3.7 Library Preparation with ONT Reagents

1. Thaw all components of the kit listed in Table 6:
2. Use 1 μL of each Fragmentation Mix RB01-96 from the provided 96-well plate to barcode each normalized sample from Subheading 3.6, **step 4** (*see* **Note 12**).
3. Mix by flicking the tube(s) or inverting the plate until homogenized, then centrifuge briefly.
4. Program a thermocycler with the following conditions in Table 7.
5. Pool all barcoded samples into a 1.5 mL Eppendorf DNA LoBind tube. Measure the final volume of pooled samples (*see* **Note 13**).
6. Vortex the pre-equilibrated room temperature AMPure XP beads well, and add an equal volume of resuspended beads to the pooled sample tube.
7. Mix well by inversion.
8. Incubate the tube at room temperature for 5 min with occasional inverting.
9. Briefly spin the tube to collect the contents and place it on a magnet to pellet the beads.

Table 6
Components of the kit to be thawed

Contents	Thaw condition	Mixing
Fragmentation Mix RB01-96	Not frozen—keep on ice	Pipette mix gently
Rapid Adapter (RA)	Not frozen—keep on ice	Pipette mix gently
Adapter Buffer (ADB)	Room temperature	Vortex mix
Sequencing Buffer (SB)	Room temperature	Vortex mix
Library Beads (LIB)	Room temperature	Vortex mix immediately before use
Flow Cell Flush (FCF)	Room temperature	Pipette mix gently
Flow Cell Tether (FCT)	Room temperature	Pipette mix gently
Elution Buffer (EB)	Room temperature	Pipette mix
AMPure XP beads (AMP)	Room temperature	Vortex mix immediately before use

Table 7
Thermal conditions used for amplicon fragmentation and addition of barcodes

Temperature	Time
30 °C	2 min
80 °C	2 min
4 °C	∞

10. With the tube on the magnet, carefully remove and discard the clear supernatant without disturbing the bead pellet.
11. Wash the beads with 1.5 mL of freshly prepared 80% ethanol to cover the pellet without disturbing it. Incubate for 1 min, then remove and discard the ethanol.
12. Repeat the wash step above, for a total of two washes, while keeping the tube on the magnet.
13. Spin down the tube and place it back on the magnet to pellet the beads. Remove any residual ethanol and allow the beads to dry briefly (*see* **Note 14**).
14. Remove the tube from the magnet, resuspend the pellet in 15 μL of Elution Buffer (EB), and incubate at room temperature for 10 min with occasional flicking.
15. Pellet the beads once again on the magnet until the supernatant goes clear.

16. Transfer 14 μL of the clear supernatant into a clean 1.5 mL Eppendorf DNA LoBind tube. Measure the concentration of 1 μL of the supernatant using a Qubit fluorometer (*see* Subheading 3.5, **steps 1–5**).
17. Dilute the library down to a final concentration of 100 ng/μL with Elution Buffer (EB) according to the Qubit measurement above and transfer 11 μL of the diluted library into a clean 1.5 mL Eppendorf DNA LoBind tube (*see* **Note 15**).
18. Prepare Rapid Adapter Mix by combining 0.5 μL of Rapid Adapter (RA) and 1.17 μL of Adapter Buffer (ADB).
19. Add 1 μL of the freshly prepared Rapid Adapter Mix to the 11 μL of diluted library, mix gently by flicking the tube, and incubate for 5 min at room temperature.
20. After incubation, keep this mix on ice until ready to load.

3.8 Loading the Flow Cell for Sequencing

1. Prepare flow cell priming mix in a 1.5-m microcentrifuge tube. Use the volumes listed in Table 8. Mix thoroughly by gentle pipetting (*see* **Note 16**).
2. Open the MinION device cover, slide the flow cell under the clip, and press the flow cell down to ensure proper thermal and electrical contact.
3. Open the priming port of the flow cell by sliding it clockwise, and remove air bubbles underneath the priming port by inserting the tip of a P1000 pipette set to 200 μL into the priming port and turning the wheel of a P1000 pipette to draw back until a small volume (around 10–20 μL) of buffer enters the pipette tip.
4. Load 800 μL of flow cell priming mix into the priming port without introducing air bubbles and incubate for 5 min.
5. While waiting, prepare the library mix for loading in a 1.5-mL microcentrifuge tube. Use the volumes listed in Table 9 (*see* **Note 17**).

Table 8
Components for making one reaction of flow cell priming mix

Reagent	Volume (μL)
Flow Cell Flush (FCF)	1170
Bovine Serum Albumin (BSA) at 50 mg/mL	5
Flow Cell Tether (FCT)	30
Total	1205

Table 9
Components for preparing the library mix for loading

Reagent	Volume (μL)
Sequencing buffer (SB)	37.5
Resuspended Library Beads (LIB)	25.5
Products generated in Subheading 3.7, **step 20**	12
Total	75

6. After 5 min, gently lift the SpotON sample port of a flow cell, and load 200 μL more of the flow cell priming mix into the priming port (*see* **Note 18**).
7. Pipette the prepared library up and down a few times, and immediately load 75 μL of the library mix in a dropwise fashion onto the SpotON port (*see* **Note 19**).
8. Close the SpotON sample port cover, the priming port and the MinION device lid. Ensure they are closed properly.
9. Navigate to start a run by selecting the "SQK-RBK114" as the kit type used for library preparation and "Super High Accuracy" for base calling. Leave all other settings as default.
10. Stop the run when around five million bases are sequenced for each sample (*see* **Note 20**).
11. If needed, ONT flow cells can be washed and reused after running by following the manufacturer's instructions (*see* **Note 21**).

3.9 Data Analysis

ARTIC pipeline is designed for analyzing viral nanopore sequencing data generated by using PCR-tiling amplicon methods. It involves using minimap2 to map ONT NGS reads to a reference genome, removing primer sequences from the assembled BAM file, and generating a consensus sequence genome. It is recommended for the downstream analysis here. Coverage depth of sequenced representative RSV -A and -B samples was shown in Fig. 2 with BAM files generated by the ARTIC pipeline before and after primer sequence trimming.

1. After the ONT run, navigate to the "fastq_pass" folder to find raw read data in the form of multiple fastq.gz files and transfer them to a computer with the ARTIC pipeline installed. Concatenate all fastq.gz files in each barcode folder to generate one fastq.gz file for each sample by using the "cat * fastq.gz > sample.fastq.gz" command in Linux.
2. Run the command "artic minion --normalise 0 --bed <RSVA_-primer_bed_directory> --ref

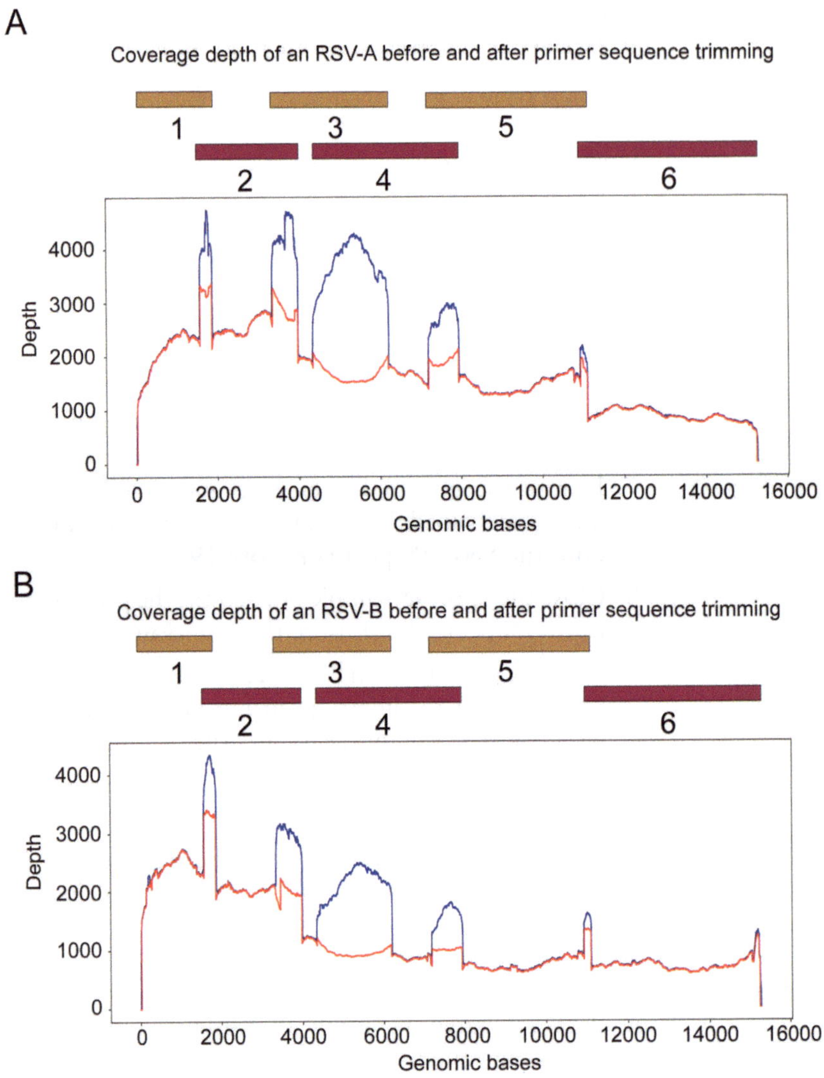

Fig. 2 Coverage depth of sequenced representative RSV-A (**a**) and RSV-B (**b**) clinical samples in genomic position covered by six overlapping amplicons [three fragments (numbered as 1, 3, and 5) colored in dark yellow were generated in pool 1 of mRT-PCR and other fragments (numbered as 2, 4, and 6) colored in dark red were generated in pool 2 of mRT-PCR] before (blue) and after (red) primer sequence trimming analyzed by ARTIC pipeline

<RSVA_reference_fasta_directory> --read-file <input_directory> <output_directory>" for mapping sample NGS data to an RSV-A reference sequence. Samples that have consensus sequences successfully generated are RSV-A samples (*see* **Note 22**).

3. Run command "artic minion --normalise 0 --bed <RSVB_primer_bed_directory> --ref <RSVB_reference_fasta_directory> --read-file <input_directory> <output_directory>" for

mapping sample NGS data to an RSV-B reference sequence. Samples that have consensus sequences successfully generated are RSV-B samples (*see* **Note 23**).

4 Notes

1. Other RNA extraction kits can be used as well, but it is recommended NOT to use total nucleic acid extraction kits because the presence of large amounts of host genome DNA can interfere with RT-PCR reactions for the production of RSV PCR fragments.
2. This aids in the removal of host DNA, which might interfere with amplicon generation and subsequent sequencing.
3. Just before use, add the required volume of 10X ezDNase Buffer and ezDNase Enzyme to a separate tube first and add 3 μL of this "master mix" to each sample, changing pipette tips between samples.
4. Do not vortex the DNase Mix.
5. Do not vortex the RT-PCR Mix.
6. RT-PCR product can be temporarily stored at 4 °C for up to 1 week, but −20 °C is recommended for longer storage as PCR products are not very stable in a fridge.
7. It is optional to visualize RT-PCR products of all samples, as the RSV whole genome can still be obtained from samples without visible RT-PCR products generated for NGS. However, it is good to check the RT-PCR products of positive controls. Apart from using TapeStation assays, other electrophoresis assays can be used as well, such as agarose gel and the E-Gel electrophoresis system (Thermo Fisher).
8. Ensure there are no bubbles present at the bottom of the tube or plate.
9. The concentration of RT-PCR products can also be measured in other ways, such as using QuantiFluor dsDNA System (Promega) on a fluorescence plate reader and TapeStation assays. Quantifying RT-PCR product with Qubit fluorometer, which is good for a small number of samples, but when handling a lot of samples, it is better to do that by using QuantiFluor dsDNA System on a fluorescence plate reader. The concentration results come along with RT-PCR product visualization analysis on a TapeStation instrument can be used as well. However, the concentration of multiplex RT-PCR products measured on a TapeStation instrument is normally underestimated.
10. It is better to prepare more than five samples for each run; otherwise, it is very difficult to get a high enough amount of libraries for flow cell loading.

11. If the volume of normalized RT-PCR products is more than 10 μL, do enrichment by adjusting the volume of normalized RT-PCR products with nuclease-free water to 50 μL and then mixing well with 50 μL of suspended and pre-equilibrated room temperature AMPure XP Beads (Beckman Coulter) supplied in the kits. Incubate for 5 min at room temperature. Spin down the sample and pellet the beads on a magnet until the supernatant is clear. Leave the tubes or plate on the magnet, pipette off the clear supernatant, and wash the beads twice with 1.5 mL of freshly prepared 80% ethanol without disturbing them. Remove the ethanol by using a pipette. Briefly spin down and place the tubes or plate back on the magnet. Pipette off any residual ethanol. Air-dry the bead pellet on the magnet for 30 s. Do not over-dry the bead pellet. Remove the tubes or plate from the magnetic plate, and add 11 μL of water to each sample. Mix by pipetting and incubate at room temperature for 2 min. Place the tubes or plate on the magnet until the beads form a pellet. Transfer 10 μL of the clear supernatant containing the DNA to a clean thin-walled PCR 0.2-mL strip tube or well of a 96-well thermal cycling plate without disturbing the bead pellet.
12. Ensure that the plate of Fragmentation Mix RB01-96 is properly and tightly sealed after use to avoid cross-contaminating Fragmentation Mix.
13. If processing more than 48 samples, pool samples into two 1.5 mL Eppendorf DNA LoBind tubes as the final volume will not fit into one tube.
14. Take care not to over-dry the pellet until it cracks as this significantly affects elution yield.
15. It is all right to load less than 100 ng/μL, but using less than 30 ng/μL may result in a very small portion of nanopores being used for sequencing, making sequencing time much longer, and making it difficult to get enough data.
16. Avoid the formation of bubbles, as bubbles can kill nanopores on the flow cells.
17. Ensure the LIB beads are well mixed to ensure homogeneity as they settle quickly.
18. The SpotON port will pulse as the priming mix is added to the priming port. Ensure liquid doesn't leave the SpotON port by dispensing slowly.
19. Each drop will be drawn into the flow cell by the vacuum created.
20. How to roughly calculate the average depth by bases. Assume the total length of the RSV genome is 15,200 bases, then the average depth is the number of bases collected by the

sequencing run divided by 15,200. For example, if five million bases are collected for one sample, then the average depth is around 329. That is enough to confidently call a consensus sequence for this sample.

21. Avoid loading different samples barcoded with the same Fragmentation Mix to the same flow cell, even in different runs after flow cell washing, as this can make the signals from the previous sample carried over to the later runs for samples with the same barcode.
22. <RSVA_primer_bed_directory> is the path to the directory of RSVA.primer.bed file, <RSVA_reference_fasta_directory> is the path to the directory of RSVA.reference.fasta file, <input_-directory> is the path to the directory of concatenated fastq.gz files, and < output_directory> is the path to the directory of analysis result.
23. <RSVB_primer_bed_directory> is the path to the directory of RSVB.primer.bed file, <RSVB_reference_fasta_directory> is the path to the directory of RSVB.reference.fasta file, <input_-directory> is the path to the directory of concatenated fastq.gz files, and < output_directory> is the path to the directory of analysis result.

Acknowledgments

The WHO Collaborating Centre for Reference and Research on Influenza is supported by the Australian Government Department of Health.

Conflict of Interest Statement The authors have no conflicts of interest to declare that are relevant to the content of this chapter.

References

1. Pandya MC, Callahan SM, Savchenko KG, Stobart CC (2019) A contemporary view of respiratory syncytial virus (RSV) biology and strain-specific differences. Pathogens 8(2)
2. Zhou X, Jiang M, Wang F, Qian Y, Song Q, Sun Y et al (2022) Immune escaping of the novel genotypes of human respiratory syncytial virus based on gene sequence variation. Front Immunol 13:1084139
3. Fourati S, Reslan A, Bourret J, Casalegno JS, Rahou Y, Chollet L et al (2025) Genotypic and phenotypic characterisation of respiratory syncytial virus after nirsevimab breakthrough infections: a large, multicentre, observational, real-world study. Lancet Infect Dis 25(3): 301–311
4. Dong X, Deng YM, Aziz A, Whitney P, Clark J, Harris P et al (2023) A simplified, amplicon-based method for whole genome sequencing of human respiratory syncytial viruses. J Clin Virol 161:105423
5. Dong X, Edwards S, Deng YM, Dapat C, Hirankitti A, Wordsworth R et al (2025) An improved rapid and sensitive long amplicon method for Nanopore-based RSV whole-genome sequencing. Influenza Other Respir Viruses 19(5):e70106

Chapter 12

Laser Microdissection and Near Single-Cell Whole Genome Amplification

Roberto Cruz-Flores, Jorge Cáceres-Martínez, and Arun K. Dhar

Abstract

Laser microdissection (LMD) and whole genome amplification (WGA) are powerful techniques that integrate molecular and histological approaches to enable the precise selection of a minimal number of virus-infected cells—down to near single-cell resolution—and the subsequent generation of whole viral genomes with minimal host DNA interference. This chapter presents a detailed protocol for LMD and near single-cell WGA, specifically optimized for the recovery and sequencing of viral genomes from formalin-fixed paraffin-embedded (FFPE) tissues. The method allows for the targeted isolation of infected cells, thereby reducing host genomic background and enhancing the detection of pathogen-specific signals for downstream next-generation sequencing (NGS). The protocol includes steps for tissue section preparation, cell isolation via LMD, DNA extraction using the PicoPure DNA Extraction Kit, and unbiased genome amplification using the SeqPlex DNA Amplification Kit—ensuring high-quality nucleic acid recovery suitable for NGS workflows.

Key words Laser microdissection, Whole genome amplification, FFPE, Single-cell genomics, Pathogen detection, Metagenomics, Shrimp viruses, SeqPlex

1 Introduction

Histopathological diagnostics played a foundational role in the development of specific pathogen free (SPF) shrimp lines during the early 1990s, revolutionizing shrimp farming by enabling more effective disease surveillance and health management [1]. Throughout the 1990s and 2000s, however, advances in molecular techniques—particularly polymerase chain reaction (PCR) and quantitative PCR (qPCR)—gradually supplanted histological methods as the gold standard for pathogen detection in aquaculture [1]. These molecular tools offer high sensitivity, specificity, and throughput, but they are fundamentally limited by their dependence on sequence complementarity. As a result, these techniques fail to detect pathogens with sequence variations in primer- or probe-binding regions, an increasingly common issue given the

Yafeng Ma (ed.), *Whole Genome Amplification: Methods and Protocols*, Methods in Molecular Biology, vol. 3003,
https://doi.org/10.1007/978-1-0716-5060-8_12,

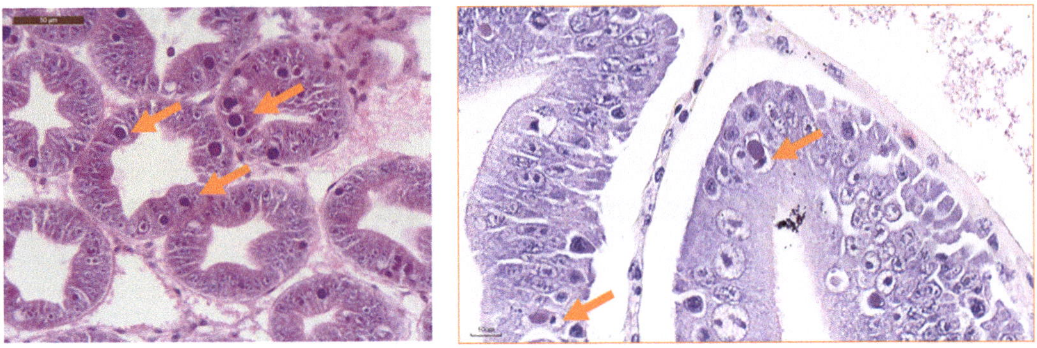

Fig. 1 Histopathology of hematoxylin and eosin (H&E)-stained sections of *Penaeus vannamei* shrimp tissue. Panel A: Intranuclear, basophilic inclusion bodies characteristics of DHPV lesions in hepatopancreatic tubule of *P. vannamei* shrimp. Panel B: Intranuclear, basophilic inclusion bodies in the midgut caecum of *P. vannamei* shrimp

high mutation rates observed in both single-stranded DNA and RNA viruses [2, 3].

Viruses such as *Decapod hepanhamaparvovirus* 1 (DHPV), a member of the *Hamaparvovirinae* subfamily, exemplify this diagnostic challenge [4]. Despite their genetic variability, these viruses consistently produce characteristic pathological changes, most notably, basophilic intranuclear inclusion bodies within hepatopancreatic epithelial cells that remain recognizable by classical histopathology even as their genome evolves. This stability of histopathological presentation stands in stark contrast to the dynamic nature of viral nucleotide sequences, underscoring the need to combining histopathology along with molecular diagnostics for accurately identifying emerging as well as existing pathogens.

In 2021, we encountered a case that epitomized these limitations: histological analysis of *Penaeus vannamei* samples from Latin America revealed pathognomonic DHPV lesions (Fig. 1), yet the samples tested negative using all available PCR, qPCR, and in situ hybridization protocols [4]. Suspecting a divergent viral genotype, we implemented a novel approach to characterize the causative agent (Fig. 2). Using laser microdissection (LMD), we selectively isolated infected cells containing inclusion bodies, thereby minimizing host DNA contamination and maximizing pathogen signal in the genome sequence derived from an infected cell. Given the minute quantity of nucleic acids retrieved from these microdissected cells, we employed whole genome amplification (WGA) to increase DNA yield for downstream sequencing.

The strategy proved very successful in identifying the newly emerging DHPV strain. The high-throughput sequencing of the amplified DNA enabled us to reconstruct the complete genome of a previously unrecognized DHPV variant, which we designated

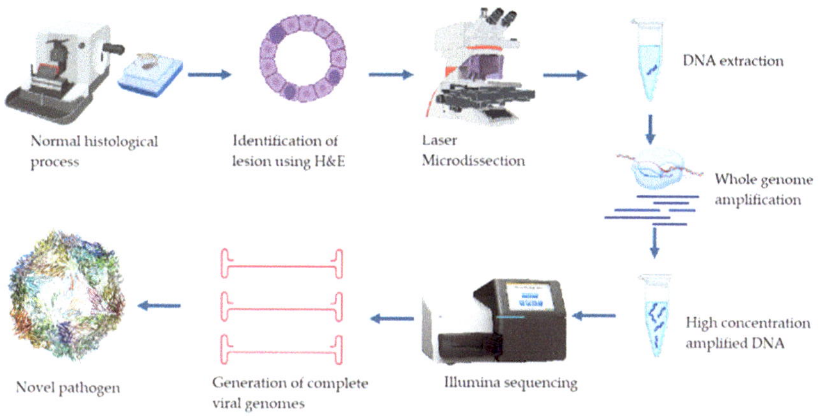

Fig. 2 Pathogen discovery pipeline: Combining histopathology, laser microdissection (LMD), and next generation sequencing (NGS) to identify a novel strain of *Decapod hepanhamaparvovirus* 1 (DHPV). (Reproduced from Cruz-Flores and Dhar [4])

genotype V. This innovative approach not only resolved the diagnostic impasse but also demonstrated the utility of LMD and WGA in characterizing novel viral pathogens from formalin-fixed paraffin-embedded (FFPE) tissues, diagnostic biological samples that are typically considered incompatible with a high-quality molecular analysis.

Laser microdissection is a powerful tool for isolating specific cells or small regions of tissue from morphologically complex or heterogeneous samples [5, 6]. When coupled with whole genome amplification, it enables genomic analyses from extremely low-input or degraded DNA such as those that are derived from rare, infected cells in archival FFPE blocks. In this chapter, we provide a detailed protocol that combines LMD, PicoPure DNA extraction, and the SeqPlex DNA Amplification Kit, offering a highly effective method for recovering pathogen genomes from FFPE tissues. This workflow is particularly valuable for retrospective investigations, pathogen discovery, and the identification of emerging viral genotypes and can be readily applied in diagnostic laboratories and reference centers engaged in basic and applied research in aquatic animals and aquaculture biosecurity.

2 Materials

2.1 Equipment

Laser Microdissection Microscope (e.g., Leica LMD7).

Microtome for FFPE tissue sectioning.

Incubators (37 °C, 55 °C, 95 °C).

Thermal cycler (for DNA amplification).

NanoDrop.

Vortex mixer and microcentrifuge.

Microscope slides with polyethylene naphthalate (PEN) membrane (Life Technologies).

0.2 mL or 0.5 mL thin-walled PCR tubes.

2.2 Reagents

2.2.1 Tissue Processing and Staining

Xylene (for deparaffinization).

Ethanol (100%, 90%, 70%).

Arcturus® Paradise PLUS Reagent System (for LMD staining).

PicoPure DNA Extraction Kit (Thermo Fisher).

RNase-free water.

2.2.2 Whole Genome Amplification (WGA)

SeqPlex Enhanced DNA Amplification Kit (Sigma-Aldrich).

Library Preparation Buffer (LP100).

Library Preparation Enzyme (E0531).

5x Amplification Mix (A5112).

DNA Polymerase for SeqPlex (SP300).

10x Primer Removal Buffer (SR401).

Primer Removal Enzyme (SR402).

Molecular Biology Grade Water (W4502).

GenElute™ PCR Clean-Up Kit (for purification).

3 Methods

Carry out all procedures at room temperature unless otherwise specified.

3.1 FFPE Tissue Preparation

- Before use, drain the tissue flotation water bath, rinse it with sterile distilled water, clean it with 95% ethanol, and refill it with fresh sterile distilled water. For the rotary microtome, remove and discard the old disposable blades. Clean the knife holder and adjacent areas first with RNase AWAY solution to eliminate potential contaminants, followed by a rinse with 95% ethanol to ensure complete sterilization. This step is very important to eliminate extraneous sequences that are not related to the sample under examination (*see* **Note 1**).
- Place 5 μm FFPE sections on polyethylene naphthalate (PEN) membrane slides.
- Bake the slides in the oven at 65 °C (10 min).
- Fill each labeled plastic slide jar with 25 mL of the corresponding solution, ensuring complete submersion of the slides. These jars are provided with the Arcturus® Paradise PLUS Reagent System.
- Immerse slides in xylene (2 min in each jar). Invert the jar carefully.

Rehydrate slides in graded ethanol series:

- 100% ethanol (1 min).
- 95% ethanol (1 min).
- 75% ethanol (1 min).
- Rinse in nuclease-free water (30 s).

Staining with Paradise® PLUS Staining Solution:

- Remove the slides and place them on a sterile paper towel. Using an RNase-free pipette tip, apply 100 μL of Paradise® PLUS Staining Solution evenly over the tissue section (*see* **Notes 2** and **3**). Gently tilt or move the slide to ensure complete coverage and uniform staining across the entire tissue area.
- Incubate at room temperature for 30 s. The slide should display a uniform blue tone. The presence of whitish areas within the tissue indicates uneven application of the Paradise® PLUS Staining Solution.
- Tap off excess stain before proceeding.

Dehydration:
Transfer slides sequentially to:

- 75% ethanol (1 min).
- 95% ethanol (1 min).
- 100% ethanol (1 min).
- Immerse slides in xylene (10 min).

Drying:

- Place slides on sterile paper towels in a fume hood to air-dry for 10 min. Properly dried slides will exhibit a uniform baby blue hue across the entire tissue section.
- Perform LCM within 2 h after removal from xylene.

3.2 Laser Microdissection (LMD)

- Place prepared slides onto the microscope stage.
- Utilize the Leica LMD7 system to selectively isolate three to five cells containing intranuclear inclusion bodies, which are pathognomonic of viral infection by DHPV. These inclusion bodies exhibit a distinct bright blue to azure coloration, contrasting with the surrounding tissue's grayish to brownish hues (Fig. 3) (*see* **Note 4**).
- Microdissect cells directly into sterile 0.6 mL microcentrifuge tube caps and immediately close the tube.

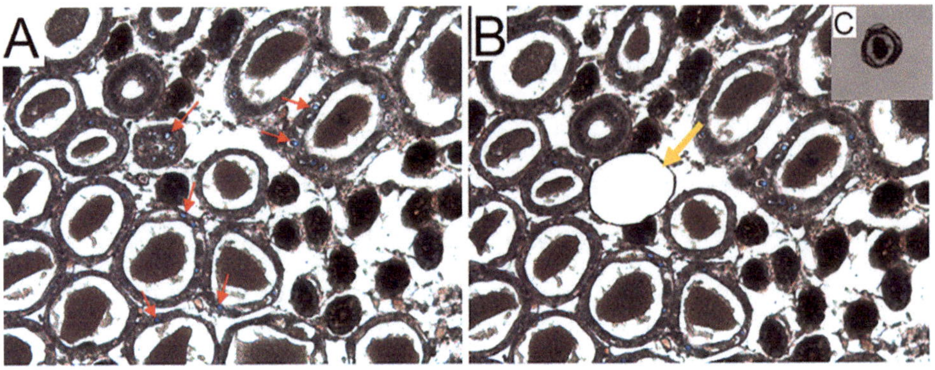

Fig. 3 Dissection of cells displaying DHPV lesions using a LMD. Panel A: Hepatopancreas tissue section mounted on PEN slides stained with Paradise Plus. The DHPV-like inclusions stain light greenish. Panel B: Same section where a heavily dissected tube was cut out by LMD. Insert: Dissected tubule containing cells ready for DNA extraction. (Reproduced from Cruz-Flores and Dhar [4])

3.3 DNA Extraction from Microdissected Cells Using Arcturus PicoPure DNA Extraction Kit

- Centrifuge 0.6 mL microcentrifuge tubes containing the microdissected cells at 14,000 RPM for 5 min using a benchtop centrifuge.
- Add 50 μL PicoPure DNA Extraction Buffer to each tube.
- Incubate at 65 °C for overnight (~18 h).
- Heat at 95 °C for 10 min (to inactivate enzymes).
- Store the DNA at −20 °C or proceed to whole genome amplification.

3.4 Whole Genome Amplification (SeqPlex Kit)

3.4.1 Pre-amplification

- Thaw the Library Preparation Buffer (LP100) and 5× Amplification Mix (A5112), and keep them on ice until use to maintain reagent stability.
- In a sterile PCR tube, prepare a 14 μL reaction by combining 2 μL of Library Preparation Buffer with 12 μL of extracted DNA.
- Mix, centrifuge briefly, and incubate in a thermal cycler: 95 °C for 2 min then on 4 °C hold.
- After the sample has cooled to 4 °C, add 1 μL of Library Preparation Enzyme (E0531). Cap tube and mix thoroughly. Centrifuge briefly and immediately proceed to the next step.
- Place reaction in a thermal cycler and incubate as follows:

  16 °C for 20 min.

  24 °C for 20 min.

  37 °C for 20 min.

  75 °C for 5 min.

  4 °C hold.
- Remove sample from thermal cycler and centrifuge briefly. Proceed to the amplification step immediately.

3.4.2 Amplification

- Add the following reagents to the 15 μL of pre-amplification product from the previous step; the total volume is 75 μL (*see* **Note 5**):

  15 μL 5X Amplification Mix (A5112).

  1.5 μL DNA Polymerase for SeqPlex (SP300).

  43.5 μL Water (W4502).

- Immediately place the mix in the thermal cycler with the following program:

  94 °C for 2 min.

  28 cycles of:

  94 °C for 15 s.

  70 °C for 5 min.

  70 °C for 30 min.

  4 °C hold.

3.4.3 Purification

- Insert a GenElute™ mini spin column (with blue O-ring) into a 2 mL collection tube.
- Add 500 μL of Column Preparation Solution to the column. Centrifuge at 12,000 × rpm for 1 min. Discard the flow-through.
- Add 400 μL of binding solution to 1 volume of amplified DNA.
- For example, add 500 μL of binding solution to 100 μL of PCR product.
- Mix thoroughly and transfer the mixture to the prepared column.
- Centrifuge at 12,000–16,000× g for 1 min. Discard the flow-through and return the column to the collection tube.
- Add 500 μL of wash solution to the column. Centrifuge at maximum speed for 1 min. Discard the flow-through and retain the collection tube.
- To remove residual ethanol, centrifuge the column without any additional solution at maximum speed for 2 min. Discard the collection tube and any remaining flow-through.
- Transfer the column to a clean 2 mL microcentrifuge tube. Add 50 μL of Elution Solution or nuclease-free water directly to the center of the column membrane.
- Incubate at room temperature for 1 min. Elute the DNA by centrifugation at maximum speed for 1 min.
- Quantify the whole genome-amplified viral DNA—originating from near single-cell input—using a NanoDrop™ 2000

spectrophotometer. The expected DNA concentration should range between approximately 50 and 75 ng/μL.

- If the DNA concentration falls within this range, the whole genome-amplified viral DNA is ready to proceed to the primer removal step.

3.4.4 Primer Removal

- On ice, assemble the primer removal reaction in a sterile, RNase/DNase-free PCR tube as follows (total volume 76.25 μL): 8.0 μL of 10× Primer Removal Buffer (SR401), 1.6 μL of Primer Removal Solution (SR400), 48 μL of purified SeqPlex viral DNA (50–75 ng/μL), 19 μL of nuclease-free water.
- Mix the reaction gently.
- Add 3.75 μL of Primer Removal Enzyme (SR402).
- Incubate the reactions (± enzyme) in a thermal cycler under the following conditions: 37 °C for 60 min, 65 °C for 20 min, hold at 4 °C.
- Upon completion, briefly centrifuge and mix the reactions.
- Immediately purify the remaining primer removal reaction using the GenElute PCR Clean-Up Kit as previously described in Subheading 3.4.3. DNA concentration will range from 20 ng/μL to 30 ng/μL.
- The whole genome-amplified DNA, obtained from near-single-cell laser microdissected tissue, is now suitable for downstream analysis via next-generation sequencing using Illumina platforms.

4 Notes

1. Decontamination of the microtome prior to sectioning FFPE tissue blocks is essential to minimize the risk of introducing exogenous DNA contamination from environmental sources. This step is particularly critical in high-throughput diagnostic laboratories, where the volume of samples increases the likelihood of cross-contamination.
2. In some cases, the tissue area on PEN membrane slides may be large enough that 100 μL of Paradise® PLUS Staining Solution is insufficient for uniform coverage. In such instances, the volume can be increased to up to 200 μL to ensure complete and consistent staining.
3. Staining tissue section with hematoxylin and eosin (H&E) stain is not compatible for downstream genomic work. This is why tissue sections are stained with paradise plus stain.

4. Careful selection of cells containing clearly identifiable viral inclusion bodies is essential. Because viral genome reconstruction typically relies on just one or a few isolated cells, only those with unequivocal evidence of viral inclusions should be selected to maximize the likelihood of successful genome recovery.
5. Monitoring the amplification of LMD-derived DNA can, if desired, be performed using a real-time thermocycler with SYBR Green-based fluorescence detection. This can be achieved using the following alternative reaction mixture:

   15 μL 5X Amplification Mix (A5112).

   1.5 μL DNA Polymerase for SeqPlex (SP300)

   1 μL SYBR Green I (S9430) diluted 1/1000.

   42.5 μL water (W4502).

References

1. Dhar AK, Pantoja-Morales CR (2024) Reestablishing histopathology as an essential component of health assessment for farmed shrimp in the era of molecular diagnostics. Rev Fish Sci Aquac 1:1–9. https://doi.org/10.1080/23308249.2024.2401583
2. Duffy S (2018) Why are RNA virus mutation rates so damn high? PLoS Biol 16:e3000003. https://doi.org/10.1371/journal.pbio.3000003
3. Robles-Sikisaka R, Bohonak AJ, McClenaghan LR, Dhar AK (2010) Genetic signature of rapid IHHNV (infectious hypodermal and hematopoietic necrosis virus) expansion in wild Penaeus shrimp populations. PLoS One 5:e11799. https://doi.org/10.1371/journal.pone.0011799
4. Cruz-Flores R, Dhar AK (2024) Tracking the emergence of a novel genotype of decapod hepanhamaparvovirus in shrimp using laser microdissection and next generation sequencing. PLoS One 19:e0311592. https://doi.org/10.1371/journal.pone.0311592
5. Murray GI (2018) Laser capture microdissection methods in molecular biology 1723
6. Cruz-Flores R, López-Carvallo JA, Cáceres-Martínez J, Dhar AK (2022) Microbiome analysis from formalin-fixed paraffin-embedded tissues: current challenges and future perspectives. J Microbiol Methods 196:106476. https://doi.org/10.1016/j.mimet.2022.106476

Chapter 13

Whole Genome Amplification of Microbial DNA from Host-Depleted Clinical Samples

Carl Halford, Lluis Moragues-Solanas, Simon A. Weller, and Matthew Gilmour

Abstract

Clinical metagenomics (CMg) involves the untargeted sequencing of the genetic content of samples collected from patients and is a highly promising method for the diagnosis of infectious disease. Depending on the sample type, CMg can be reliant on the removal of the host genetic material from the sample to support detection of microbial pathogens, and this selective process (or an otherwise low abundance of microbial cells in the sample) may result in concentrations of DNA too low for productive sequencing. Whole genome amplification (WGA), the nonselective amplification of the total DNA of a sample, can be applied to significantly increase the concentration of DNA and enable CMg sequencing. This chapter describes the methods for the amplification of microbial DNA extracted from host-depleted wound swab samples using the GenomiPhi™ V3 Ready-To-Go™ (Cytiva) DNA WGA kit and host-depleted whole blood samples using the REPLI-g® Single-Cell WGA kit (Qiagen). This is followed by the de-branching and bead-based clean-up of the amplified DNA, resulting in highly concentrated DNA ready for CMg DNA sequencing.

Key words Clinical metagenomics, Host-depletion, Whole genome amplification, De-branching

1 Introduction

1.1 Clinical Metagenomics

CMg has the potential to provide rapid, untargeted infectious disease diagnostics for a range of clinical sample types, such as tissues, fluids, and wounds [1, 2]. The application of CMg for diagnostics has progressed to the stage where workflows are being more routinely implemented within diagnostic laboratories for the rapid identification of pathogens and the provision of additional clinically relevant information [3, 4]. However, there are several limitations associated with CMg, including the high proportion of host genetic material relative to the microbial species and the low total DNA yields following extraction, particularly in samples that have undergone host-depletion steps, resulting in poor pathogen

Yafeng Ma (ed.), *Whole Genome Amplification: Methods and Protocols*, Methods in Molecular Biology, vol. 3003, https://doi.org/10.1007/978-1-0716-5060-8_13,

detection sensitivity. The application of WGA can significantly increase the concentration of total DNA from these low biomass samples, allowing for downstream processing and CMg methods to be applied to sample types that would otherwise not be possible. This chapter describes two WGA methods, using two different commercial kits, developed for the amplification of microbial DNA obtained from host-depleted clinical samples, such as wound swab samples [2] and whole blood [5].

1.2 Whole Genome Amplification for Clinical Metagenomics

WGA refers to the nonselective amplification of the total DNA of a sample, resulting in a significant increase in the total DNA concentration, reflective of the entire genetic content of the sample. This enables the detection and study of DNA obtained from very low biomass samples, including trace DNA and single cells [6].

There are several amplification methods available for WGA, including polymerase chain reaction-based, isothermal, and hybrid amplification methods, as previously reviewed [7]. Isothermal amplification methods of WGA include multiple displacement amplification (MDA), the most commonly applied WGA method. MDA uses a DNA polymerase with high extension activity and random primers to amplify the genome using isothermal conditions. Advantages of this method include the amplification of the whole genome with low error rates [8] and enabling fragments of 10–100 kilobases in length to be amplified with high fidelity [9]. In addition, the inclusion of the WGA step can increase the DNA concentration from as low as >0.05 ng/μL up to <100 ng/μL, allowing a wider range of detection methods to be carried out [2, 5].

Despite being required for the generation of sufficient DNA from low biomass samples, there are disadvantages to applying WGA methods. Firstly, the amplification stage adds 2–3 h to the total workflow and turnaround time, varying according to kit selection. Also, WGA can introduce biases to the sequencing data through the preferential amplification of certain DNA sequences compared to others, according to factors such as GC content and DNA conformation [10]. Therefore, it may be necessary to better understand and account for the impacts of these biases on the detection of pathogen species prior to selecting a suitable WGA method. Finally, the amplification of DNA by MDA forms branched intermediate DNA structures and chimeras [11] that can interfere with downstream processes, including blocking the protein nanopores used for nanopore-based DNA sequencing. Therefore, a DNA debranching step, involving the incubation of the DNA with an endonuclease to break up the branched structures, is required. This reduces the average read length to two to three kilobases [12] and adds to the overall method complexity.

The methods described in this chapter were developed for the amplification of microbial DNA extracted from host-depleted wound swab samples using the GenomiPhi V3 Ready-To-Go DNA amplification kit and host-depleted whole blood samples using the REPLI-g Single Cell kit [2, 5]. Each method has advantages and disadvantages that are suited to the sample type and the aims of the diagnostic test, and these factors should be considered before selecting a WGA method for a specific application (*see* **Note 4.1**). Both the GenomiPhi V3 and REPLI-g Single Cell Kits carry out WGA from single cells and low-DNA-input samples using an MDA approach with isothermal amplification by the Phi29 polymerase. The input DNA is denatured, and random hexamer primers anneal to the template strand, followed by strand displacement synthesis. Additional priming and synthesis occurs on the displaced strands, resulting in the production of large amounts of highly branched DNA (>1 μg per reaction) with long fragment sizes (2–100 kb) in a few hours [13, 14].

Using these methods, we have produced high concentrations of microbial DNA from low biomass, host-depleted clinical samples, ready for nanopore-based DNA sequencing for the detection of the causative pathogen of infectious disease.

2 Materials

2.1 GenomiPhi V3 Ready-to-Go Whole Genome Amplification

- Up to 10 ng input DNA, made up to 10 μL total volume with nuclease-free water.
- GenomiPhi V3 Ready-To-Go DNA amplification kit, including:
  - 2 × denaturation buffer.
  - Freeze-dried reagent cake within 0.2 mL PCR tube.
  - PCR tube caps.
- Mini microcentrifuge.
- PCR thermocycler.
- Ice box.

2.2 Qiagen REPLI-g Single Cell Whole Genome Amplification

- ≥50 fg input DNA, made up to 15 μL total volume with nuclease-free water.
- PCR thermocycler.
- Thick-walled 0.2 mL PCR tube.
- REPLI-g Single Cell Kit amplification kit, including:
  - DLB buffer (reconstituted by adding 500 μL of nuclease-free water).
  - DL-Dithiothreitol 1 M (DTT).
  - Stop solution.

– Single-cell DNA polymerase.
– Single-cell reaction buffer.

2.3 DNA Quantification

- Amplified DNA.
- Qubit™ 4 Fluorometer (Invitrogen).
- Qubit dsDNA HS Assay Kit (Invitrogen), including:
  – Qubit dsDNA HS reagent.
  – Qubit dsDNA HS buffer.
  – Qubit dsDNA HS Standard 1.
  – Qubit dsDNA HS Standard 2.
- Qubit assay tubes (Invitrogen).

2.4 DNA De-Branching and Bead Clean-Up

- Up to 1 μg of amplified DNA sample made up to a total volume of 17 μL with nuclease-free water.
- T7 Endonuclease I (New England Biolabs).
- NEBuffer™ 2 (New England Biolabs).
- PCR thermocycler.
- Thick-walled 0.2 mL PCR tube.
- AMPure XP bead-based reagent (Beckman Coulter).
- DynaMag-2™ magnetic rack (Invitrogen).
- 1.5 mL DNA LoBind tubes (Eppendorf).
- Molecular biology grade ethanol, absolute.
- Nuclease-free water.

3 Methods

3.1 GenomiPhi V3 Ready-to-Go Whole Genome Amplification

1. When amplifying DNA from sample types containing a higher input DNA amount (≥ 10 ng) or where DNA from multiple microbial species is expected to be present, the GenomiPhi V3 kit can be selected for the WGA step (*see* **Notes 4.1.1** and **4.2.1**).
2. Working within a laminar flow hood, ideally situated within a PCR-clean room (*see* **Note 4.2.2**), add 10 ng of input DNA, made up to 10 μL with nuclease-free water, to a thick-walled 0.2 mL PCR tube.
3. Include a negative control sample at this stage, made up of nuclease-free water, to be processed simultaneously, to monitor for contamination (*see* **Note 4.2.3**).
4. Add 10 μL of 2 × denaturation buffer to the sample and mix thoroughly by pipetting.

5. Using a PCR thermocycler, heat the sample to 95 °C for 3 min. Remove the sample, and place it on ice for at least 1 min, to cool the sample to approximately 4 °C.
6. Transfer 20 μL of the denatured DNA sample to the 0.2 mL PCR tube containing the freeze-dried reagent cake. Mix by pipetting until the reagent cake is dissolved, and close the tube using the caps provided (*see* **Note 4.2.4**). Using a bench-top mini microcentrifuge, spin the tubes and ensure the liquid is collected at the bottom of the tube.
7. Using a PCR thermocycler, incubate the sample at 30 °C for 2 h, followed by 65 °C for 10 min (*see* **Note 4.2.5**).
8. Remove the sample from the PCR thermocycler, and add it to ice to cool the sample to 4 °C.

3.2 REPLI-g Single Cell Whole Genome Amplification

1. When amplifying DNA from samples containing very low concentrations of DNA (<10 ng), the REPLI-g Single Cell WGA kit can be selected.
2. Add ≥50 fg of input DNA in a total volume of 15 μL with nuclease-free water, in a thick-walled 0.2 mL PCR tube.
3. Prepare fresh D2 buffer for each sample to be mixed with the template DNA. For each sample, mix 0.16 μL of DTT with 1.84 μL of DLB buffer.
4. Add 2 μL of D2 buffer to 15 μL of DNA template. Vortex briefly and spin down. Incubate the sample tubes at 65 °C for 10 min.
5. After the incubation has completed, add 3 μL of Stop Solution to each sample at room temperature. Vortex briefly and spin down. Incubate for 5 min at room temperature, then place the samples on ice while preparing the polymerase master mix.
6. Prepare the polymerase master mix by adding 29 μL of single-cell reaction buffer and 2 μL of single-cell DNA polymerase per sample. Then, add 31 μL of the master mix to each sample (already containing 20 μL). Vortex briefly and spin down.
7. Incubate the samples at 37 °C for 90 min, followed by heat inactivation at 60 °C for 3 min.
8. The samples can then be taken forward immediately for processing or stored at −20 °C.

3.3 DNA Quantification

1. Remove the Qubit dsDNA HS Assay Kit reagents from storage and keep at room temperature until all reagents have reached room temperature.
2. Set up one Qubit assay tube per sample to be quantified, plus two additional tubes for the DNA standards. Label the top of each tube.

3. Prepare a master mix of the Qubit working solution. Add 199 μL of Qubit dsDNA HS buffer per sample plus standards to be measured to a clean 2.0 mL plastic tube. Add 1 μL Qubit dsDNA HS reagent per sample plus standards to be measured. Mix by pulse-vortexing the tube for 10 s.
4. To the two standard Qubit assay tubes, add 190 μL of working solution. Pulse-vortex the Qubit dsDNA HS Standard 1 for 5 s and add 10 μL to Standard Tube 1. Repeat for Qubit dsDNA HS Standard 2 and add to Standard Tube 2. Store the standards at room temperature in the dark whilst the remaining samples are prepared.
5. Add 198 μL of the Qubit working solution to each Qubit assay tube for sample quantification. Add 2 μL of the sample to each tube. Pulse-vortex each Qubit assay tube, including the standards, for 5 s. Incubate at room temperature in the dark for 2 min.
6. On the qubit 4 Fluorometer screen, select dsDNA, dsDNA assay, and Read Standards.
7. Insert Standard 1 into the sample chamber, close the lid, and select Read Standard.
8. Repeat for Standard 2.
9. Select Run Samples. Select the sample volume of 2 μL and the output units as ng/μL.
10. Insert the Qubit assay tube into the sample chamber and select Read tube. Record the concentration of DNA in each sample (*see* **Note 4.4.1**).

3.4 DNA De-Branching and Bead Clean-Up

1. In a clean plastic tube, prepare the de-branching master mix by adding 2 μL NEBuffer 2 and 1 μL T7 Endonuclease I and mix by pipetting (*see* **Note 4.5.1**). Prepare sufficient master mix for 1 additional reaction, to account for pipetting errors (*see* **Note 4.5.2**).
2. Transfer up to 1 μg of amplified DNA made up to a total volume of 17 μL with nuclease-free water to a new, clean 0.2 mL PCR tube (*see* **Note 4.5.3**). Add 3 μL of the de-branching master mix and mix by pipetting. Spin the tube down on a benchtop mini microcentrifuge and incubate the tubes on a PCR thermocycler at 37 °C for 15 min. Following the incubation step, add the tubes to ice to cool to 4 °C.
3. Remove the AMPure XP beads from the fridge, and store at room temperature for at least 15 min prior to use (*see* **Note 4.5.4**).
4. Transfer the de-branched DNA to a 1.5 mL low-bind tube, and mix with 12 μL AMPure XP beads, and incubate for 5 min at room temperature on a slow shaking hula mixer using the

following settings: rotation speed 15 and timer 10, reciprocal tilting turning angle 45° and timer 5, vibration turning angle 5° and timer 5.

5. Place the tube on a DynaMag-2 magnetic rack, wait until the solution is clear and the beads have pelleted to the side of the tube (~3–5 min), and pipette away the supernatant.
6. With the tube still on the magnetic rack, add 200 μL of freshly prepared 70% ethanol. Incubate the tube at room temperature for 30 s before removing the supernatant.
7. Repeat the previous step. Remove any residual ethanol by pulse-spinning the tube, placing the tube back onto the magnetic rack and pipetting off any remaining 70% ethanol.
8. Open the tube lid and air-dry the pellet for 2 min at room temperature outside the magnet, ensuring that the pellet does not dry to the point of cracking.
9. Resuspend the beads in 15 μL of nuclease-free water by pipetting up and down, and incubate the tube at room temperature for 2 min off the magnetic rack.
10. Place the tube back on the magnetic rack and wait until the solution is clear (~2 min). Transfer the supernatant, avoiding transferring any of the beads, to a new 1.5 mL DNA Lo-bind tube and dispose of the remaining beads.
11. The de-branched DNA can then be stored at −20 °C or progressed directly to downstream processing, including library preparation and DNA sequencing. The library preparation and DNA sequencing method selected for the next steps will depend on the aims of the experiment. The methods described in this chapter were developed for nanopore-based DNA sequencing, chosen for its suitability for use in austere environments. Regarding the library preparation method, when prioritizing reduced method complexity and result turnaround time, the Rapid Barcoding kit may be selected due to the reduced total number of steps. Alternatively, when aiming to achieve increased sequencing depth, read length and data quality, the ligation sequencing kit may be selected.

4 Notes

4.1 Selection of Whole Genome Amplification Kits

1. The concentration and composition of the DNA sample will vary, depending upon a range of factors, including the clinical sample type, the inclusion of a host depletion step, and the initial microbial load. From a diagnostics perspective, certain sample types, including wound and respiratory samples, are likely to contain multiple microbial species, reflective of the more diverse microbiomes at these sites. In these cases, the

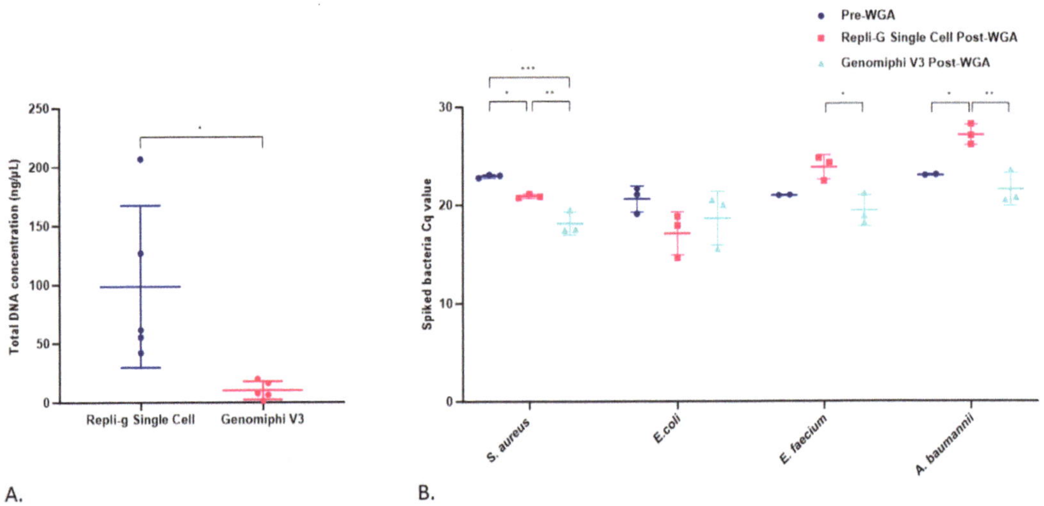

Fig. 1 Comparisons of the total DNA concentrations (**a**) and the amplification of bacterial species DNA (**b**) by the REPLI-g Single Cell and GenomiPhi V3 WGA kits.

proportion of sequencing reads assigned to each species represents key clinical information and may be indicative of the causative pathogen of infection. Therefore, it is important that the WGA step does not introduce significant biases to the sequencing data through the overamplification of certain species' DNA. In contrast, sample types such as blood will contain lower total microbial DNA content originating from fewer microbial species. Therefore, increasing the total DNA yield may be more of a priority. The selection of an appropriate WGA kit will be informed by these factors (*see* Methods 3.1 and 3.2).

Previous evaluations of WGA kits in which purified bacterial DNA was amplified by the two WGA kits indicated that the GenomiPhi V3 and REPLI-G kits are better suited for the amplification of different sample types. As demonstrated by Fig. 1b, when comparing the amplification of purified bacterial DNA from multiple species, the GenomiPhi V3 Ready-To-Go kit demonstrated improved DNA amplification across the four tested species, resulting in lower Cq values for three of the four bacterial species. Therefore, this kit may be considered to be more suitable for the amplification of nucleic acids extracted from samples such as wound or respiratory samples, where the uniform amplification of DNA from different species is required. However, the total DNA yield obtained from the REPLI-g Single-Cell kit was significantly higher (Fig. 1a). Therefore, the REPLI-g kit may be considered to be more suited to the amplification of bacterial DNA from sample types such as blood, in which the total microbial DNA content is lower and fewer microbial species are likely to be present.

Amplifying low concentrations of purified bacterial DNA, the REPLI-g single cell kit resulted in higher total DNA yields (1A). However, the GenomiPhi V3 kit demonstrated improved amplification of the range of bacterial species DNA, measured by real-time PCRspiepr146 (1B).

4.2 GenomiPhi V3 Whole Genome Amplification Kit Notes

1. The DNA extracted from clinical samples, particularly from sterile or host-depleted sample types, may contain DNA concentrations >1.0 ng/μL. In this case, alternative kits, such as the REPLI-g Single Cell kit (*see* Methods 3.2) may be preferred due to the increased DNA yield from low DNA inputs. Alternatively, >10 ng of DNA can be amplified by the GenomiPhi V3 kit (see Methods 3.1), but the total DNA yield may be significantly reduced.
2. Prior to the WGA step, metagenomic workflows are highly susceptible to contamination as contaminating sequences will be amplified, alongside the sample DNA, interfering with the interpretation of the results. It is recommended that the WGA reagents be prepared within a laminar-flow hood within a PCR-clean room to minimize the addition of contamination to the regents or the sample prior to amplification (*see* Methods 3.1–3.4).
3. The inclusion of WGA negative controls is recommended to enable the identification of contamination and aid in the determination of the contamination source, if required. Nuclease-free water is amplified alongside the DNA samples and analyzed to identify contaminating sequences (*see* Methods 3.1, **step 3**).
4. When transferring the denatured DNA sample to the reagent cake, repeated pipetting is usually required to resuspend the reagents, which must be fully dissolved prior to the incubation step. Any bubbles generated by pipetting can be removed by gently flicking the tube, followed by centrifugation to collect the sample at the bottom of the tube (*see* Methods 3.1, **step 6**).
5. The length of the amplification incubation stage should be selected depending upon the required DNA concentration of DNA for downstream processes and the concentration of the initial DNA sample. The GenomiPhi V3 manufacturer's recommendation is for a 1.5 h incubation at 30 °C, but for low biomass samples, this can be increased to 2 h or more, as we have done for the DNA extracted from host-depleted wound samples, if required (*see* Methods 3.1, **step 7**).

4.3 REPLI-g Single Cell WGA Kit Notes

1. Prior to the WGA step, metagenomic workflows are highly susceptible to contamination as contaminating sequences will be amplified alongside the sample DNA, interfering with the interpretation of the results. It is recommended that the WGA

reagents be prepared within a laminar-flow hood within a PCR-clean room to minimize the addition of contamination to the regents or the sample prior to amplification.

2. The inclusion of WGA negative controls is recommended to enable the identification of contamination and aid in the determination of the contamination source, if required. Nuclease-free water is amplified alongside the DNA samples and analyzed to identify contaminating sequences.
3. Polymerase incubation time can be extended to 3 h or more (*see* Methods 3.2, **step 6**). However, when small quantities of template (<1 ng) are used, primer dimers begin to form. After 1.5 h of incubation, a large number of primer dimers will have bound to each other, creating more complex and larger amplicons, which may also be amplified. Due to the exponential increase in primer dimers, beyond the 1.5 h time point, bacterial genomic DNA will still be amplified but at significantly lower rates.
4. Due to the high presence of primer dimers, fluorometric quantification using Qubit after the WGA protocol may overestimate the amount of bacterial genomic DNA in the sample. If precise quantification of bacterial genomic DNA is required post-WGA, a qPCR assay with generic primers for microorganism identification (e.g., 16S) is recommended.
5. This protocol has been optimized to obtain the highest amount of amplified bacterial genomic DNA in the shortest possible time. This has been achieved primarily by adding different buffers at lower concentrations (e.g., DTT for its amplification-enhancer properties [15]) and minimizing water usage to maintain the lowest reaction volume possible (*see* Methods 3.2, **steps 2** and **5**). All steps in the REPLI-g Single Cell handbook that involve buffer dilution have been omitted.
6. Whenever possible, use DNA extracts that have been cleaned and verified for purity. Poor purity ratios may affect polymerase amplification rates, potentially requiring additional incubation time.
7. Always prepare fresh D2 buffer for each experiment (*see* Methods 3.2, **step 2**). Do not store or reuse it.

4.4 DNA Quantification

1. WGA can result in highly concentrated DNA samples greater than the upper limit of quantification of the Qubit dsDNA HS Assay Kit. In these instances, dilute 2 μL of the amplified DNA in 18 μL of nuclease-free water, and repeat the quantification step (*see* Methods 3.3, **step 10**).

4.5 DNA De-Branching

1. Preparing a master mix of T7 Endonuclease I and NEBuffer 2 reduces the pipetting steps, advantageous when operating within austere environments. It must be ensured that the master mix is fully mixed by pipetting to ensure that sufficient enzyme is added to each sample (*see* Methods 3.4, **step 1**).
2. When preparing the master mix, prepare at least 1 reaction excess to account for pipetting inaccuracies when aliquoting into the sample tubes (*see* Methods 3.4, **step 1**).
3. The method states that a maximum input of 1 μg in 17 μL total volume should be de-branched for downstream processing (*see* Methods 3.4, **step 2**). However, low DNA concentration samples amplified by the GenomiPhi V3 kit may contain <1 μg DNA, even after WGA. In these cases, it is recommended that the concentration of T7 Endonuclease I enzyme is also reduced, maintaining a ratio of 10 units of enzyme per 1 μg of amplified DNA. For example, if the sample contains a total of 500 ng of DNA, the input of T7 Endonuclease I should be reduced to 5 units of enzyme, equivalent to 0.5 μL of enzyme at 10 units/μL, as supplied by the manufacturer. This prevents the over-digestion of the DNA when present at low concentrations.
4. AMPure XP beads should be equilibrated to room temperature prior to use to ensure the efficient binding of the DNA to the beads and prevent the loss of DNA (*see* Methods 3.4, **step 3**).

Conflict of Interest Statement

The authors have no conflicts of interest to declare that are relevant to the content of this chapter.

Funding

Carl Halford and Simon A. Weller were funded by the United Kingdom Ministry of Defence. The funder approved submission of the manuscript for publication, but had no role in study design, data collection and analysis, or preparation of the manuscript.
Matthew Gilmour gratefully acknowledges the support of the Biotechnology and Biological Sciences Research Council (BBSRC). Matthew Gilmour's research was funded by the BBSRC Institute Strategic Programme Microbes and Food Safety BB/X011011/1 and its constituent projects BBS/E/F/000PR13636 (Theme 3, Flexible capabilities to reduce food safety threats and respond to national needs).

References

1. Chiu CY, Miller SA (2019) Clinical metagenomics. Nat Rev Genet 20(6):341–355
2. Halford C, Le Viet T, Edge K, Russell P, Moore N, Trim F et al (2024) Development of a clinical metagenomics workflow for the diagnosis of wound infections. BMC Med Genet 17(1):276
3. Charalampous T, Alcolea-Medina A, Snell LB, Alder C, Tan M, Williams TGS et al (2024) Routine metagenomics service for ICU patients with respiratory infection. Am J Respir Crit Care Med 209(2):164–174
4. Benoit P, Brazer N, de Lorenzi-Tognon M, Kelly E, Servellita V, Oseguera M et al (2024) Seven-year performance of a clinical metagenomic next-generation sequencing test for diagnosis of central nervous system infections. Nat Med 30(12):3522–3533
5. Moragues-Solanas L, Le-Viet T, McSorley E, Halford C, Lockhart DS, Aydin A et al (2024) Development and proof-of-concept demonstration of a clinical metagenomics method for the rapid detection of bloodstream infection. BMC Med Genet 17(1):71
6. Wang X, Liu Y, Liu H, Pan W, Ren J, Zheng X et al (2022) Recent advances and application of whole genome amplification in molecular diagnosis and medicine. MedComm 3(1):e116
7. Gawad C, Koh W, Quake SR (2016) Single-cell genome sequencing: current state of the science. Nat Rev Genet 17(3):175–188
8. de Bourcy CFA, De Vlaminck I, Kanbar JN, Wang J, Gawad C, Quake SR (2014) A quantitative comparison of single-cell whole genome amplification methods. PLoS One 9(8): e105585
9. Lasken Roger S (2009) Genomic DNA amplification by the multiple displacement amplification (MDA) method. Biochem Soc Trans 37(2):450–453
10. Sabina J, Leamon JH (2015) Bias in whole genome amplification: causes and considerations. In: Whole genome amplification: Methods and protocols, pp 15–41
11. Lasken RS, Stockwell TB (2007) Mechanism of chimera formation during the multiple displacement amplification reaction. BMC Biotechnol 7:1–11
12. Dadzie FA, Beaudry MS, Deyanov A, Slanis H, Duong MQ, Turner R et al (2024) Evaluating the benefits and limits of multiple displacement amplification with whole-genome Oxford Nanopore Sequencing. bioRxiv
13. Spits C, Le Caignec C, De Rycke M, Van Haute L, Van Steirteghem A, Liebaers I et al (2006) Whole-genome multiple displacement amplification from single cells. Nat Protoc 1(4):1965–1970
14. Ahsanuddin S, Afshinnekoo E, Gandara J, Hakyemezoğlu M, Bezdan D, Minot S et al (2017) Assessment of REPLI-g multiple displacement whole genome amplification (WGA) techniques for metagenomic applications. J Biomol Tech 28(1):46
15. Ralser M, Querfurth R, Warnatz H-J, Lehrach H, Yaspo M-L, Krobitsch S (2006) An efficient and economic enhancer mix for PCR. Biochem Biophys Res Commun 347(3):747–751

Chapter 14

Whole Genomic Amplification for Detecting Single-Gene Variants Causing Inherited Disease

Tania Rojas-Pérez, Miguel Ángel Alcántara-Ortigoza, Ariadna González-del-Ángel, Felipe Camargo-Díaz, and Esther López-Bayghen

Abstract

The whole genome amplification (WGA) allows new clinical applications with minimal genetic material, such as in the genetic diagnosis of Mendelian diseases in embryos before implantation (Preimplantation Genetic Test for Mendelian Abnormalities, PGT-M). This approach allows couples to avoid the transmission of Mendelian disease by undergoing assisted reproduction treatment through in vitro fertilization (IVF). First, Preimplantation Genetic Testing for Aneuploidy (PGT-A) is used to identify chromosomal aneuploidies in IVF-generated embryos. Then, or in parallel, euploid embryos can be screened for specific diseases caused by variants in a single gene to achieve the conception of offspring free of a specific monogenic disease.

Here, we detail the WGA preparation and two downstream usages: (1) preparation of PCR fragments for Sanger sequencing, exemplifying the diseases we detected for healthy embryo selection and transfer in IVF, and (2) detection of chromosome Y for embryo sex diagnosis.

Key words WGA, PGT-A, PGT-M, Amplification, Genome, In vitro fertilization, Embryo, Monogenic disease

1 Introduction

The whole genome amplification (WGA) used mainly for preimplantation genetic diagnosis for aneuploidies (PGT-A, performed with the next-generation sequencing) also allowed the genetic diagnosis of Mendelian diseases in embryos (PGT-M) in couples who have undergone assisted reproduction treatment through fertilization in vitro (IVF). PGT-A identifies chromosomal aneuploidies in IVF-embryos, and with PGT-M, a specific disease is detected to achieve the conception of a healthy child when there is a history of monogenic disease in the family, identified by affected offspring if one of the couple has a Mendelian disease or if one or both

Yafeng Ma (ed.), *Whole Genome Amplification: Methods and Protocols*, Methods in Molecular Biology, vol. 3003, https://doi.org/10.1007/978-1-0716-5060 8_14, members of a couple are known to be carriers of a specific pathological variant [1].

Preimplantation genetic diagnosis is based on an embryo biopsy, in which a few cells are recovered (approximately 1–15) when the embryo is on day 5 or 6 of embryonic development. Since an embryonic biopsy yields few cells, WGA allows the entire genome of a cell to be amplified [2], then next-generation sequencing (NGS) methodologies are used for PGTA, and in parallel, PCR products can be amplified from the WGA to be analyzed for Sanger sequencing for detecting single-gene pathological variants inherited into the embryo genome or analyzed for identifying the Y chromosome. The Sureplex system, based on specific self-inert degenerative primers to form an in vitro molecular library utilizing flanking universal priming sites, is a suitable template to amplify PCR fragments from 100 to 700 bp. According to the company brochures, an allelic dropout rate limited to 10% can be expected, which significantly improves over previous PCR-based methods [3].

We have been exploring the use of WGA for the diagnosis of different monogenic diseases, allowing embryo selection and obtaining satisfactory results; the process starts by taking 1–15 cells obtained through an embryo biopsy, amplifying the genomic DNA by endpoint PCR under the conditions given for the SurePlex™ DNA Amplification System kit and then the monogenic analysis for the identification. Although genetic testing of one or a few cells is challenging and the overall procedure is quite complex, WGA allows an increased number of genome copies by polymerase chain reaction amplification from nanogram amounts of starting material into micrograms, with minimal null alleles [4]. Genetic tests can be offered for all disorders where the genetic pathological variants are known, including their inheritance pattern (autosomal vs. X-linked or dominant vs. recessive). In this study, 14 monogenic diseases were identified (Table 1) out of 14 couples, demonstrating that a preceding WGA step provided sufficient DNA to assess the embryo's genetic abnormalities using PCR-based techniques and Sanger sequencing or for qPCR (Fig. 1) to identify the Y chromosome and assess the sex (Fig. 2) [5].

2 Materials

2.1 Equipment

Microcentrifuge (Eppendorf).

Centrifuge (Thermo Fisher Scientific).

Microplate shaker (BioShake XP).

Endpoint thermal cycler (Bio-Rad).

Analytical balance (Ohaus).

Table 1
Diagnosis of genetic alterations in monogenic hereditary diseases before embryo transfer in IVF

Disease	Gene	Form of inheritance	Analyzed variant inherited from mother	Analyzed variant inherited from father	Affected embryos/ observations	Reference
Pendred syndrome and autosomal recessive non-syndromic hearing loss, type 4 (DFNB4)	*SLC26A4*	Autosomal recessive	c.85G > C o p. (Glu29Gln)	c.707 T > C o p. (Leu236Pro)	Compound heterozygote	[10]
Zellweger syndrome	*PEX2*	Autosomal recessive	c.782A > G o p. (His261Arg)	c.782A > G o p. (His261Arg)	Homozygote	[11]
Cystic fibrosis	*CFTR*	Autosomal recessive	c.1210-34TG[13]T [5]	c.1210-34TG[12]T [5]	Homozygous 5 T is usually asymptomatic.	[12]
Syndrome of usher type II (USH2)	*USH2A*	Autosomal recessive	c.2276G > T o p. (Cys759Phe)	c.9433C > T o p. (Leu3145Phe	Compound heterozygote	[13]
Rhizomelic chondrodysplasia punctata type 1	*PEX7*	Autosomal recessive	c.289_290del o p. (Thr97Cysfs*11)	c.289_290del o p. (Thr97Cysfs*11)	Homozygote Pair with two entities by variants in *PEX7* and *G6PD*.	[14]
Alpha 1 antitrypsin deficiency	*SERPINA1*	Autosomal recessive	c.863A > T o p. (Glu288Val)	c.1178C > T o p. (Pro393Leu)	Compound heterozygote	[15]
Acidemia propionic	*PCCA*	Autosomal recessive	c.2041-1G > T o p. (?)	c.2041-1G > T o p.(?)	Homozygote	[16, 17]
Sickle cell anemia	*HBB*	Autosomal recessive	c.19G > A o p. (Glu7Lys)	c.19G > A o p. (Glu7Lys)	Homozygote	[18]
Polycystic kidney disease in adults	*PKD1*	Autosomal dominant	NA	c.1549_1553 delinsCCCTGTGCT o p.(Cys517Profs*71)	Heterozygote	[19]

(continued)

Table 1 (continued)

Disease	Gene	Form of inheritance	Analyzed variant inherited from mother	Analyzed variant inherited from father	Affected embryos/ observations	Reference
Spinocerebellar ataxia type 1 (SCA1)	*ATXN1*	Autosomal dominant	22 y 23 repeated normal range	>44 CAG repeats abnormal expansion range	Heterozygote	[15]
Diamond-Blackfan anemia type 6 (DBA6)	*RPL5*	Autosomal dominant	NA	NA	Heterozygote de novo. A child with variant c.175_176del or p.[Asp59Tyrfs*53] de novo; parents without the variant	[20]
Osteogenesis imperfect	*CO1A1*	Autosomal dominant	NA	c.1177C > T o p. (Gln393Ter)	Heterozygote	[21]
Insensitivity to androgens	*AR*	X-linked	c.2318A > G o p. (Glu773Gly)	NA	Affected occasional male/ heterozygous female hemizygous	[22]
Glucose-6-phosphate dehydrogenase deficiency	*G6PD*	X-linked	c.563C > T or p. (Ser188Phe) / c.563C > T or p. (Ser188Phe)	c.563C > T or p. (Ser188Phe)	Male hemizygote/female homozygous	[14]
Hemophilia A	*F8*	X-linked	c.788-2A > G o p. (?)	NA	Hemizygous males/ heterozygous females are occasionally affected	[23]

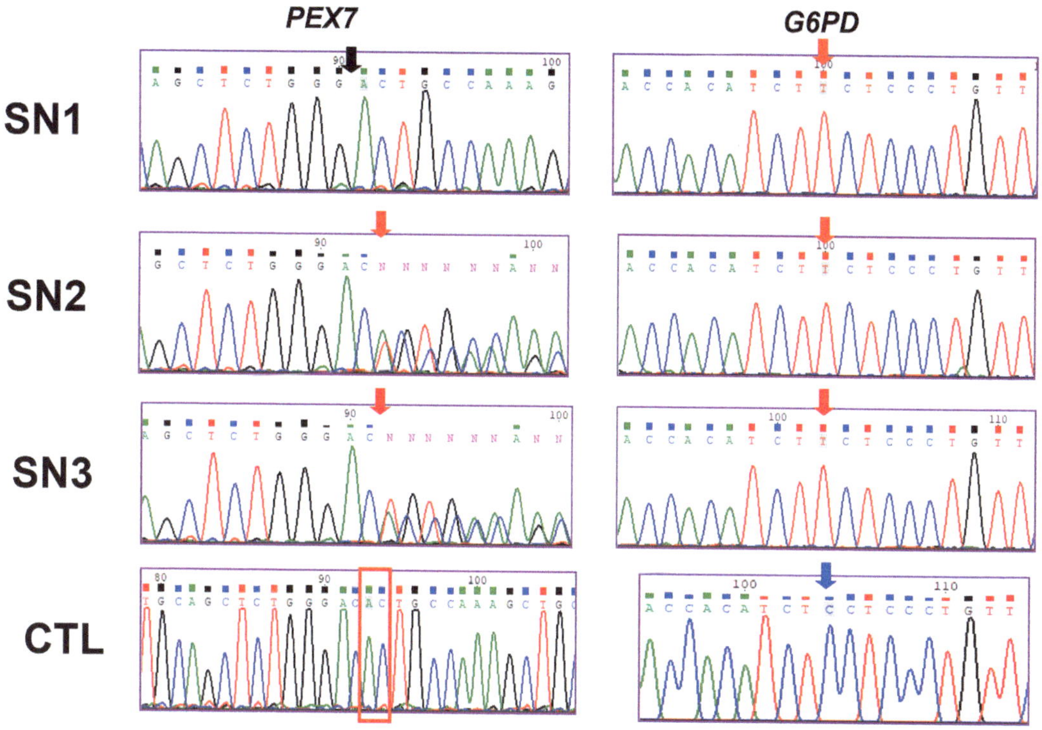

Fig. 1 Partial electropherograms (forward strand) of exon 3 and 6 of the *PEX7* and *G6PD* genes, respectively. The red arrow in *PEX7* electropherograms indicates the position where the overlapping of the reads derived from the normal allele and those derived from the NM_000288.4:c.289_290del or p.(Thr97Cysfs*11), which indicates a heterozygous *PEX7* genotype in embryos SN2 and SN3. The black arrow in *PEX7* electropherograms of SN1 indicates the homozygous (affected) genotype for the deletion of the dinucleotide "AC" (typical sequence: 5′-...GGGACACTGC...-3′; mutated sequence: 5′-...GGGACTGC...-3′) or NM_000288.4: c.289_290del or p.(Thr97Cysfs*11). The open square indicates the presence of the normal dinucleotide "AC" at positions c.289 and c.290 of the *PEX7* gene in the standard control (CTL). Instead, the red arrow in *G6PD* electropherograms indicates the homozygous or hemizygous genotypes for the variant NM_001360016.2:c.563C > T or p.(Ser188Phe). The blue arrow reveals the regular *G6PD* "C" allele for the c.563 positions in the healthy female control (CTL). Only one electropherogram of the three obtained for embryos SN1, SN2, and SN3 is shown; no genotype discrepancy was observed

Horizontal electrophoresis chamber for agarose gels (Danger).

Power supply for electrophoresis (Danger).

PCR hood (Esco).

Epoch microplate spectrophotometer (BioTek).

MixMate for PCR tubes and plates (Eppendorf).

Picofuge (Labnet).

StepOnePlus Thermal Cycler (Applied Biosystems).

Thermoblock (LabTech).

MultiDoc-It Transilluminator with UV Light, Orange Filter (UVP).

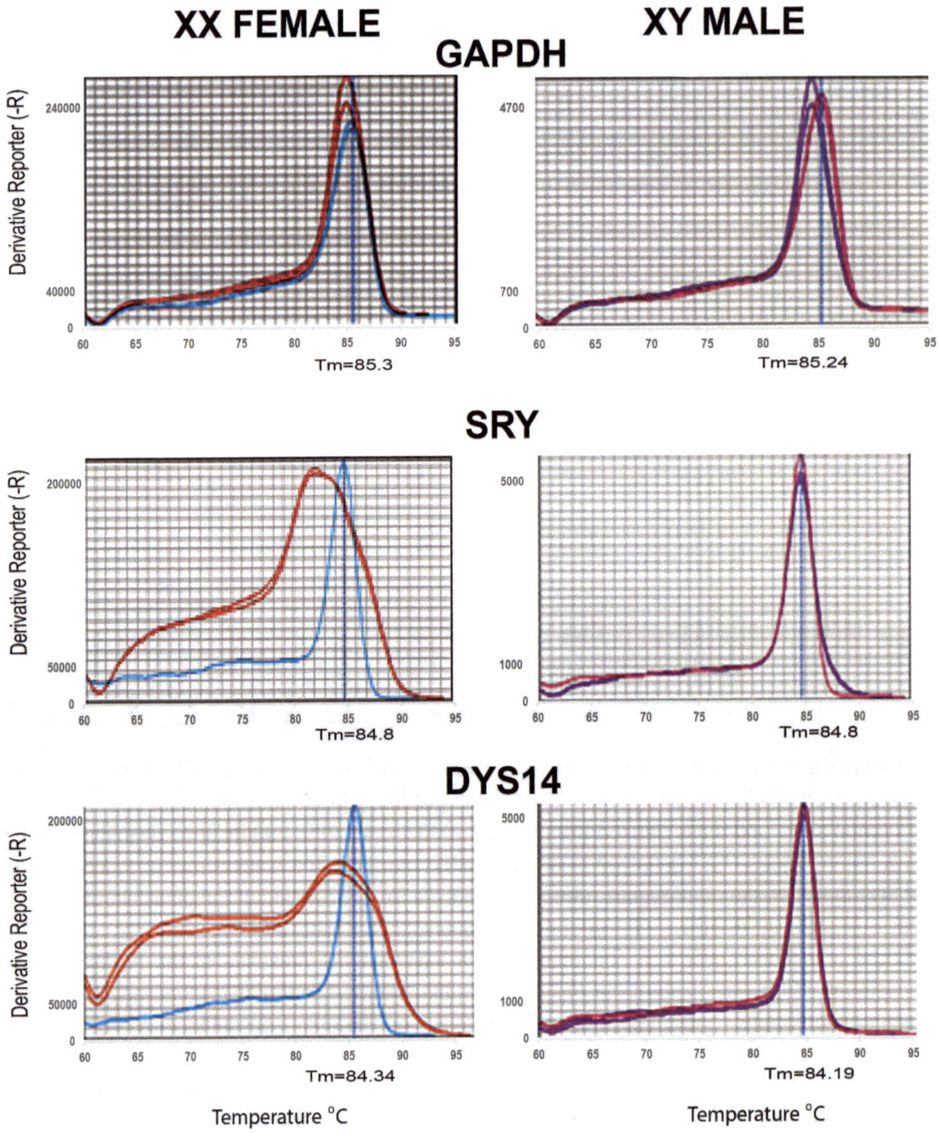

Fig. 2 GAPDH, SRY, and DYS14 melting curve analysis. WGA from a blastomere was used as a template for GAPDH-, SRY-, and DYS14-specific qPCR reactions. Representative curves for males and females are shown. Melting temperature peaks were determined by the harmful first derivative (-dF/dT) plot. Orange and pink curves are samples; blue curves are positive controls (genomic DNA (gDNA), note 5.6). These plots were used to identify the embryos' sex. Notice no amplification in the orange curves on the left side for XX female WGA

Refrigerated centrifuge Biofuge Primo; jackets for 1 mL and 600 μL tubes.

Spatula for agarose gels.

Microtube racks.

Cooling racks for PCR tubes.

1000, 200, 10, and 2 μL micropipettes.

1000 and 600 μL microtubes (Axygen).

1000, 200, 10, and 2 μL micropipette tips (Axygen).

Hazardous waste container.

Applied Biosystems PCR tubes and caps (0.2 mL).

2.2 Whole Genome Amplification (WGA)

SurePlex™ DNA Amplification System Illumina Kit (Illumina, 15053626).

Cell Extraction: Dilution buffer and enzyme for cell extraction.

Pre-amplification: Pre-amplification buffer and pre-amplification enzyme.

Amplification: Amplification buffer, amplification enzyme, and nuclease-free water.

2.3 Electrophoresis

1X TBE Solution: Prepare with 10X TBE Electrophoresis Buffer (Thermo Scientific B52: Tris base [tris(hydroxymethyl) aminomethane], boric acid, and ethylenediaminetetraacetic acid (EDTA), and mix the solution manually.

1% agarose (Vivantis PC0701–5): Weigh 3 grams of agarose and transfer it to a 500 mL Erlenmeyer flask. Add 300 mL of 1X TBE and microwave for 2 minutes until the powder dissolves; store at room temperature.

Loading buffer: Ethidium bromide (Sigma Aldrich 46,067), 6X DNA Loading Dye (Sigma Aldrich G2526), and water (PISA). Prepare 1 mL of buffer in an Eppendorf tube, adding 50 μL of ethidium bromide, 190 μL of 6X loading buffer, and 750 μL of water for molecular use.

2.4 PCR Amplification of a Product for Sanger Sequencing

ExoSAP-IT® PCR Product Cleanup, Affymetrix, Inc., A55242.

Big Dye® Terminator Cycle Sequencing Kit, Life Technologies.

Primer designs are carried out using Primer-BLAST software (https://www.ncbi.nlm.nih.gov/tools/primer-blast/). These must include the gene segment where the pathogenic variant is located, and PCR amplicons are intended to be 100–150 base pairs with an annealing temperature of 63–65 degrees Celsius.

2.5 PCR Amplification for Embryo Sex Determination

Pure ethanol (Sigma Aldrich 64-17-5): Molecular biology grade ethanol, prepare 50 mL aliquots.

70% ethanol: Add 35 mL of pure ethanol and 15 mL of water for injection (PISA) to a conical tube.

3 M sodium acetate solution pH 5.2(Sigma Aldrich S5636): Sodium acetate, glacial acetic acid (Sigma Aldrich 695092), and water for injection (PISA). Dissolve 12.3 g of sodium acetate (anhydrous) in ~40 mL and dilute to pH 5.2.

qPCRBIO SyGreen® Mix Hi-ROX (PCR Biosystem PB20.12-05) for real-time amplification.

Primers: GAPDH (product size: 363 bp): forward primer 5′ TTTAACTCTGGTAAAGTGGATATTG 3′ and reverse primer 5′ CTGTTGTCATACTTCTCATGGTT 3′; SRY (247 bp): forward primer 5′ GCTGGGATACCAGTGGAAAA 3′ and reverse primer 5′ TCTTGAGTGTGTGGCTTTCG-3′) and DYS14 (147 bp): forward primer 5′-CATCCAGAGCGTCCCTGG-3′ and reverse primer 5′-TTCCCCTTTGTTCCCCAAA-3′ [5].

3 Methods

3.1 Pre-WGA Steps

3.1.1 In Vitro Fertilization

All patients were subjected to controlled ovarian stimulation for 10–12 days with gonadotropin-releasing hormone (GnRH) agonists and antagonists. Briefly, the controlled ovarian stimulation protocol consisted of administering a daily dose of a GnRH antagonist (0.25 mg/day Cetrorelix, Cetrotide, Merck México, or 0.25 mg/day Ganirelix acetate, Orgalutran Laboratorio MSD) in the luteal phase after menses. Gonadotropins were administered in variable doses (with a minimal daily dose of 300 IU), dependent on the patient's age and ovarian responsiveness, with further adjustments according to serum estradiol (E_2) levels and vaginal ultrasound measurements of the follicular diameter obtained every 2 or 3 days. Stimulation was prolonged until the mean diameter of the leading follicles was >18 mm. Recombinant human chorionic gonadotropin (hCG) (Choragon 1000 IU, Ferring) was administered, and oocyte retrieval was conducted 36 h after the administration of hCG with ultrasound guidance. The partner's sperm was prepared using density gradient centrifugation. ICSI inseminated the fresh oocytes, and fertilization was judged by forming two pronuclei 19 h after insemination. Embryos were cultured in Global Total for Fertilization media (Cat # LGGT-30, LifeGlobal) and incubated at 37 °C in 8% CO_2, 5% O_2, and 87% N_2. Embryo morphological parameters evaluated were weighed into a matrix to rate each embryo, with the sum of values obtained on a scale of 0 (low quality) to 12 (high quality). Selection and embryo transfer were done according to the embryo morphological assessment, using the criteria established by the Istanbul Consensus Workshop on Embryo Assessment [6]. For Day 3 embryo biopsies (blastocysts), embryo transfers were done on Day 5 (fresh cycle); however, for Day 5 embryo biopsies (trophectoderm), the resulting blastocysts were cryopreserved by vitrification and thawed for transfer in a new natural cycle with endometrial preparation. An embryologist monitored and recorded all information about fertilization rates, embryo development, and embryo morphology for each oocyte.

3.1.2 Day 5 Embryo Biopsy

For Day 5 embryo (expanded blastocyst stage containing 50–150 cells), a laser was used to create a thin funnel in the zona pellucida on the opposite side of the inner cell mass. Blastocysts were incubated for 2–3 h to allow blastocoele expansion and herniation of the trophectoderm cells from the zone. Afterward, the embryo was placed into 20 μL of Ca^{2+}/Mg^{2+}-free bicarbonate-buffered G-PGD medium (cat #10074, Vitrolife). Applying gentle suction with the biopsy pipette (MBB-FP-SM-35, Origio, Malov, Denmark), the trophectoderm cells herniate from the zona. Four to five trophectoderm cells were dissected from each blastocyst using four laser pulses of 3-minute duration, and a total of 10–15 cells were retrieved, washed, and placed into a 0.2 μL PCR tube containing 2.5 μL of PBS (Phosphate Buffered Saline 10X, GIBCO 70011).

3.2 Detailed Method for WGA

Unless otherwise specified, perform all procedures at room temperature, following the protocol provided by the DNA Amplification System (*see* **Note 5.2**).

Whole Genome Amplification (WGA): Performed from one cell to a dozen, using the endpoint polymerase chain reaction amplification with three steps:

1. Cell lysis: perform cell rupture to release the genetic material of all the cells to be amplified. By reaction, use 4.8 μL of extraction buffer and 0.2 μL of enzyme for extraction; add 5 μL of this mix to each of the samples to be processed; the process is carried out in an extraction hood, and a different tip is used for each of the samples to avoid contamination. Incubate the samples in the thermal cycler for approximately 15 min under the following conditions (Table 2):
2. Preamplification: In a 600 μL Eppendorf tube, take 4.8 μL of amplification buffer and 0.2 μL of pre-amplification enzyme, add 5 μL of volume per reaction, and add to each of the samples. Subsequently, incubate in the thermal cycler under the cycle specifications in Table 3:
3. Amplification: Once the pre-amplification is finished, keep the samples cold; add (per reaction) 25 μL of amplification buffer, 0.8 μL of amplification enzyme, and 34.2 μL of RNAse-free water. Incubate the samples in the thermal cycler under the conditions in Table 4 (approximately 43 min):

Electrophoresis: Verify the amplification of the complete genome by 1% agarose electrophoresis; if no other procedure is used, store the remaining sample in the freezer at −20 °C (it can be stored for 1 year).

Table 2
Conditions for cell lysis

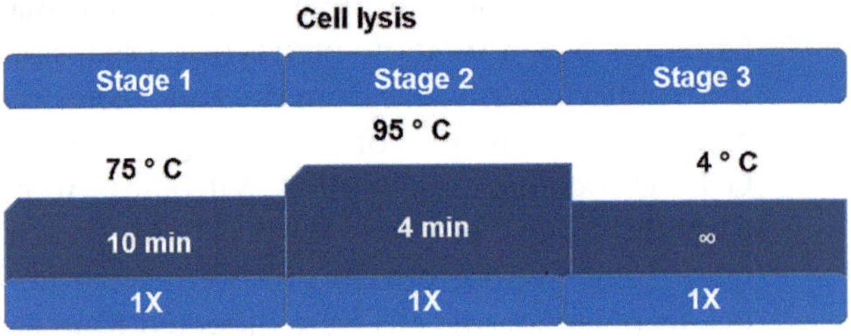

Table 3
Conditions for pre-amplification

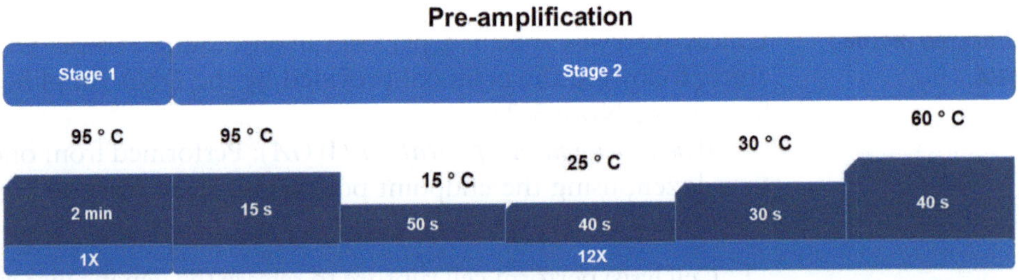

Table 4
Conditions for amplification

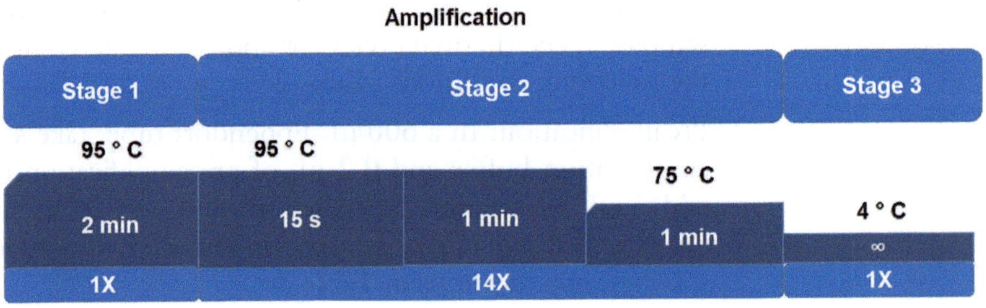

1. Heat the agarose solution in the microwave oven until dissolved, then pour it into the electrophoresis mold, place the corresponding combs, and let it stand until the gel solidifies. Once the gel is solid, remove the comb and transfer it to the electrophoresis chamber. Pour TBE 1x until the gel is completely covered.
2. For each sample, prepare a drop of 3 μL of loading buffer, load 10 μL of WGA, mix by pipetting, and place in a gel well; repeat this step for each sample (*see* **Note 5.1** and **5.3**).

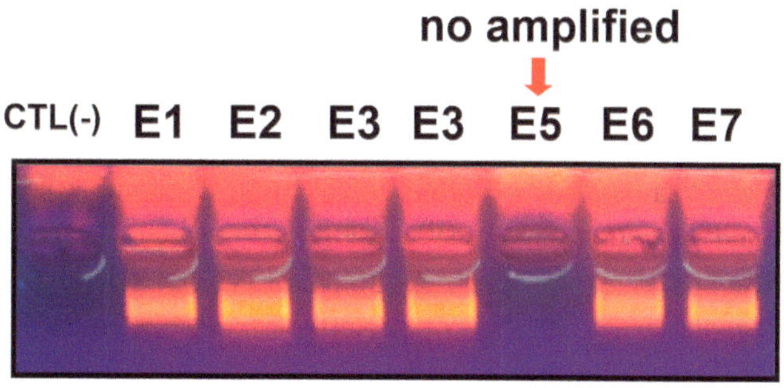

Fig. 3 *Verification of the correct WGA amplification. Agarose gel image. E1–6 refers to embryoEmbryos number*

3. Close the electrophoresis chamber, turn on the power source, and run the gel at 100 volts for 7 min.
4. Remove the gel from the chamber and place it in the UV transilluminator to visualize the amplification (Fig. 3).

3.3 Downstream Application: PCR Amplification and Sanger Sequencing

1. PCR verification: PCR products must be of excellent quality and quantity, free of other inhibitors, primer dimers, or non-specific products. Under conventional PCR conditions involving 1X of reaction buffer, 2.5 mM of magnesium chloride, 1 UI of Taq DNA polymerase, and 800 nM of each primer, the specific amplicons are generated under the following universal PCR program: 95 °C 5 min, followed by 30 cycles at 95 °C 30 s, 63–65 °C 30 s, and 72 °C 50 s, followed with a single extension step at 72 °C, 10 min, and finish with a 4 °C (infinite).
2. The amplicons are evaluated for quality and quantity by electrophoresis in agarose gels before purification by an enzymatic method.
3. Subsequent analysis is performed by directly automated sequencing of the "forward" strand (Sanger type with Big Dye® Terminator Cycle Sequencing kit, Life Technologies) and alignment with the reference sequences.
4. The unpurified PCR product generated can be sent to Psomagen Inc. (https://order.psomagen.com/). The amplicons are enzymatically purified before the BigDye terminator reaction (ExoSAP-IT PCR Product Cleanup Reagent, Thermo Fisher A55242).
5. Amplicon's identity was verified by sequencing using BigDye Terminator v3.1 reagent and the Reverse primer (Applied Biosystems, Foster City, CA, USA). Samples were sequenced with the ABI PRISM 3700 analyzer (Applied Biosystems), and sequences were subsequently aligned using ChromasPro

version 2.1.10 with the reference genomic and mRNA sequences. The variants found are annotated using Alamut Visual Plus version 1.11.

3.4 Downstream PCR Application: Real-Time PCR for Sex Determination

3.4.1 WGA Purification

1. In a newly labeled 600 μL tube, 4 μL of acetic acid and 400 μL of absolute ethanol (*see* **Note 5.4**).
2. Add 20 μL of WGA—vortex for 5 s to homogenize the sample.
3. Incubate at −72 °C in an ultra-freezer for 15 min.
4. Centrifuge in the refrigerated centrifuge at 4 °C with a speed of 13,000 g for 15 min.
5. Decant the supernatant, keeping only the pellet (which is white at the bottom of the tube). Add 200 μL of 70% ethanol and vortex for 5 s.
6. Centrifuge in a refrigerated ultracentrifuge at 4 °C with a power of 13,000 g for 5 min.
7. Decant the supernatant, then add another 200 μL of 70% ethanol and vortex for 5 s—centrifuge in a refrigerated ultracentrifuge at 4 °C with a power of 13,000 g for 5 min.
8. Decant the supernatant and perform a 3-second spin in the PicoF.
9. uge to remove all the remaining ethanol from the tube.
10. Using a 10 μL micropipette, carefully remove the remaining supernatant (ethanol) until only the white pellet remains.
11. Incubate the opened tubes in a thermoblock at 54 °C for 2 min to evaporate the remaining ethanol.
12. Hydrate the pellet with 100 μL of nuclease-free water. Vortex for 5 s to homogenize the sample, then spin for 10 s (*see* **Note 5.5**).

3.4.2 Real-Time qPCR Using WGA as a Template

1. Determine the number of reactions and the arrangement of samples by primers, as well as positive (genomic DNA, *see* **Note 5.6**) and negative (no template) controls. Determine the plate design according to the annealing temperature for each gene.
2. Master mix preparation: The genes to be evaluated are *GAPDH, SRY*, and *DYS14* [7–9]. In 600 μL tubes previously labeled with the name of each primer, the qPCR master mix is 5 μL; the Forward Primer 0.2 μL (20pMol); the reverse primer 0.2 μL (20pMol); and the molecular quality water is 3.6 μL. Vortex each mix for 5 s and place them in the cold chain. Arrange the tubes in the PCR plate and add 9 μL of each mix according to the primer name. Use different tips to avoid contamination. Add 1 μL of water to each negative control

tube, and replace the caps to prevent contamination of other samples. Add 1 μL of WGA template to each tube according to the PCR plate design, and cap the tubes as you finish loading a complete sample row. Use separate tips to avoid contamination between samples. Add 1 μL of genomic DNA (positive control) to each tube as appropriate and cap. The centrifuge in the PCR tube PicoFuge (0.1–0.2 mL) is used for 3 s to sink all samples to the bottom of the tubes.

3.4.3 Programming StepOnePlus (Real-Time PCR)

1. Turn on the device connected to the computer that contains the StepOnePlus program. Once the program is open, select Advanced Setup in the Setup section. Enter the name of the experiment where Untitled 1 appears.
2. Select the 96-well instrument, Ct quantification, SYBR Green Reagents, and Fast (around 40 min to complete the run) options in the Experiment Properties section.
3. Select the Plate Setup section. Select the Add New Target option in the left section to add the gene names.
4. Add the names of the samples to be analyzed by selecting the Add New Sample option in the right section.
5. Select the Assign Targets and Samples tab.
6. Complete according to the experimental design plate, assigning the samples and genes to each well.
7. Select the option: Enable VeriFlex™Block so that the plate is divided into six sections and the equipment allows gene analysis at different annealing temperatures.
8. Select the boxes assigning the corresponding gene according to the established plan, and select the negative controls for each gene used.
9. Select the section that says Run Method. Change the reaction volume from 20 μL to 10 μL.
10. Set the following thermal cycling conditions: Start 42 °C—5 min, 95 °C—5 min; Denaturation 95 °C—15 sec, alignment 56 °C, 58 °C, 58 °C, 58 °C, 62 °C, and 62 °C—3 sec (temperatures by zone), elongation 72 °C—1 min (reading), and melting curve: 95 °C—15 sec, 60 °C - 1 min, 95 °C—15 sec. Cycles (30). Insert the plate into StepOnePlus; the letter A on the plate should be facing the upper left corner—press START RUN.
11. Once the experiment is complete, access the Analysis submenu to view the results; a split screen will appear. On the left side, you can see the amplification curves obtained (Fig. 2), and on the right side, the plate layout designed in the program; in the drawing of each well, you can see the CT value.

4 Notes

1. Because of ethidium bromide, the gel is disposed of in a hazardous waste container with nitrile gloves.
2. To avoid contamination, latex gloves must be worn at all times, and an 80% concentrated antibenzyl solution must be used to clean work surfaces.
3. To maintain its effectiveness, prepare the loading buffer in 1 mL aliquots and amber tubes (Axygen).
4. Keep the prepared ethanol solutions refrigerated at 4 °C.
5. The samples should be quantified after WGA purification, considering an optimal concentration of 100–150 ng/μL and purity evaluated via a 260/280 ratio ≥1.8 and a 260/230 ratio ≥1.8.
6. The genomic DNA used as a positive control in the PCR can be extracted from peripheral blood using GenElute TM Blood Genomic DNA (Sigma NA2010-1KT).

Acknowledgments

We thank Leticia Ramírez-Martínez for her technical expertise.

Conflict of Interest Statement The authors have no conflicts of interest to declare that are relevant to the content of this chapter.

References

1. Volozonoka L, Miskova A, Gailite L (2022) Whole genome amplification in preimplantation genetic testing in the era of massively parallel sequencing. Int J Mol Sci 23
2. Zheng YM, Wang N, Li L, Jin F (2011) Whole genome amplification in preimplantation genetic diagnosis. J Zhejiang Univ Sci B 12: 1–11
3. Deleye L, Coninck DD, Christodoulou C, Sante T, Dheedene A, Heindryckx B et al (2015) Whole genome amplification with SurePlex results in better copy number alteration detection using sequencing data compared to the MALBAC method. Sci Rep 5:11711
4. Estévez Gómez N (2018) Whole-genome amplification of cancer and healthy human single cells. Universidade de Vigo (España), Vigo (España)
5. Schaeffer E, Porchia LM, Neumann A, Luna A, Rojas T, López-Bayghen E (2020) Embryos derived from donor or patient oocytes are not different for in vitro fertilization outcomes when PGT allows euploid embryo selection: a retrospective study. Clin Transl Med 9:1–10
6. Embryology ASiRMaESIGo (2011) The Istanbul consensus workshop on embryo assessment: proceedings of an expert meeting. Hum Reprod 26:1270–1283
7. Zargari M, Sadeghi MR, Shahhosseiny MH, Kamali K, Saliminejad K, Esmaeilzadeh A et al (2011) Fetal sex determination using non-invasive method of cell-free fetal DNA in maternal plasma of pregnant women during 6 (th)–10(th) weeks of gestation. Avicenna J Med Biotechnol 3:201–206
8. Cui KH, Warnes GM, Jeffrey R, Matthews CD (1994) Sex determination of preimplantation embryos by human testis-determining-gene amplification. Lancet 343:79–82

9. Schaeffer E, López-Bayghen B, Neumann A, Porchia LM, Camacho R, Garrido E et al (2017) Whole genome amplification of day 3 or day 5 human embryos biopsies provides a suitable DNA template for PCR-based techniques for genotyping, a complement of preimplantation genetic testing. Biomed Res Int 2017:1209158
10. Smith RJH, Azaiez H, Odell AM (1993–2025) SLC26A4-related sensorineural hearing loss, In: Adam MP, Feldman J, Mirzaa GM, Pagon RA, Wallace SE, Amemiya a, editors. GeneReviews(®) [Internet]. Seattle (WA): University of Washington, Seattle. https://www.ncbi.nlm.nih.gov/books/NBK1467/
11. Ebberink MS, Mooijer PA, Gootjes J, Koster J, Wanders RJ, Waterham HR (2011) Genetic classification and mutational spectrum of more than 600 patients with a Zellweger syndrome spectrum disorder. Hum Mutat 32:59–69
12. Savant A, Lyman B, Bojanowski C, Upadia J (1993–2025) Cystic fibrosis, in: Adam MP, Feldman J, Mirzaa GM, Pagon RA, Wallace SE, Amemiya A (eds). GeneReviews(®) [Internet]. Seattle (WA): University of Washington, Seattle. https://www.ncbi.nlm.nih.gov/books/NBK1250/
13. Koenekoop R, Arriaga M, Trzupek KM, Lentz J (1993–2025) Usher syndrome type II, In: Adam MP, Feldman J, Mirzaa GM, Pagon RA, Wallace SE, Amemiya A (eds). GeneReviews(®) [Internet]. Seattle (WA): University of Washington, Seattle. https://www.ncbi.nlm.nih.gov/books/NBK1341/
14. Braverman NE, Steinberg SJ, Fallatah W, Duker A, Bober MB (1993–2025) Rhizomelic chondrodysplasia punctata type 1, in: Adam MP, Feldman J, Mirzaa GM, Pagon RA, Wallace SE, Amemiya A (eds). GeneReviews(®) İnternet. Seattle (WA): University of Washington, Seattle. https://www.ncbi.nlm.nih.gov/books/NBK1270/
15. Stoller JK, Hupertz V, Aboussouan LS (1993–2025) Alpha-1 antitrypsin deficiency Adam MP, Feldman J, Mirzaa GM, Pagon RA, Wallace SE, Amemiya A (eds) GeneReviews(®) [internet]. Seattle (WA): University of Washington, Seattle. https://www.ncbi.nlm.nih.gov/books/NBK1519/
16. Galarreta Aima CI, Shchelochkov OA, Jerves Serrano T, Venditti CP (1993-2025) Propionic Acidemia, in: Adam MP, Feldman J, Mirzaa GM, Pagon RA, Wallace SE, Amemiya A (eds). GeneReviews(®) [Internet]. Seattle (WA): University of Washington, Seattle. https://www.ncbi.nlm.nih.gov/books/NBK92946/
17. Neumann A, Alcantara-Ortigoza MA, Gonzalez-Del Angel A, Zarate Diaz NA, Santana JS, Porchia LM et al (2021) Whole-genome amplification/preimplantation genetic testing for propionic acidemia of successful pregnancy in an obligate carrier Mexican couple: a case report. World J Clin Cases 9:8797–8803
18. Nongbri SRL, Verma HK, Lakkakula B, Patra PK (2017) Presence of atypical beta globin (HBB) gene cluster haplotypes in sickle cell anemia patients of India. Rev Bras Hematol Hemoter 39:180–182
19. Wilson PD (2004) Polycystic kidney disease. N Engl J Med 350:151–164
20. Sieff C (1993-2025) Diamond-Blackfan Anemia, In: Adam MP, Feldman J, Mirzaa GM, Pagon RA, Wallace SE, Amemiya A (eds). GeneReviews(®) [Internet]. Seattle (WA): University of Washington, Seattle. https://www.ncbi.nlm.nih.gov/books/NBK7047/
21. Amu G, Zhang G, Jing N, Ma Y (2024) Developing stapled aptamers with a constrained conformation for osteogenesis imperfect therapeutics. J Med Chem 67:18883–18894
22. Singh S, Ilyayeva S (2025) Androgen insensitivity syndrome, StatPearls [Internet]. Treasure Island (FL): StatPearls Publishing. StatPearls Publishing LLC. https://www.ncbi.nlm.nih.gov/books/NBK542206/
23. Konkle BA, Nakaya Fletcher S (1993-2025) Hemophilia A, In: Adam MP, Feldman J, Mirzaa GM, Pagon RA, Wallace SE, Amemiya A (eds). GeneReviews(®) [Internet]. Seattle (WA): University of Washington, Seattle. https://www.ncbi.nlm.nih.gov/books/NBK1404/

Chapter 15

Single-Cell Low-Pass Whole Genome Sequencing of Circulating Tumor Cells

Zhuo Wang and Qihui Shi

Abstract

Single-cell whole genome sequencing (WGS) enables accurate identification and characterize circulating tumor cells (CTCs) in blood and non-blood body fluids, leading to a non- or minimally invasive liquid biopsy approach for cancer diagnosis and prognosis. Here, we describe a single-cell low-pass WGS protocol for sensitive and accurate CTCs detection in blood and non-blood body fluids by combining a metabolic function-associated marker and a Tn5 transposome-based WGS method.

Key words Circulating tumor cells, Single-cell sequencing, Copy number alteration, Whole genome sequencing

1 Introduction

The current gold standard for cancer diagnosis is based on the microscopic examination of a tissue sample taken from the site of a suspected cancer ("biopsy") by a pathologist. As an invasive procedure, tumor biopsy is particularly challenging in cancers with accessibility or safety concerns, repeated sampling, and asymptomatic or early-stage cancers [1]. Body fluids (e.g., urine, blood, pleural effusion, ascites, cerebrospinal fluid, bile) have been reported as possible surrogates to tumor tissues because they contain disseminated tumor cells, nucleic acids, or exosomes from tumor sites [2]. The liquid biopsy represents a non- or minimally invasive means for clinicians to interrogate cancers [2]. Compared with molecular markers, the detection of intact tumor cells in body fluids provides compelling evidence of malignancy [1–3]. To date, cytologic examination based on morphological features is the only diagnostic method to identify tumor cells in body fluids [4, 5]. Although high specificity in diagnosis, cytology suffers from low diagnostic sensitivity due to the rarity of malignant cells in body fluids and substantial overlapping cytologic features

Yafeng Ma (ed.), *Whole Genome Amplification: Methods and Protocols*, Methods in Molecular Biology, vol. 3003,
https://doi.org/10.1007/978-1-0716-5060-8_15,

between malignant and non-malignant cells [4, 5]. Meanwhile, current methods for detecting circulating tumor cells (CTCs) in peripheral blood are based on epithelial markers that suffer from low sensitivity and limited specificity and cannot be applied to non-blood body fluids that contain a large number of benign cells of epithelial origin (e.g., urine, pleural effusion, ascites) [2, 3].

Here, we describe a single-cell low-pass whole genome sequencing (WGS)-based method to accurately detect and characterize CTCs in blood and non-blood body fluids for establishing definitive cancer diagnosis across different cancer types [4–6]. This liquid biopsy-based, multicancer diagnostic method combines sensitive detection of suspicious CTCs with a metabolic function-associated marker and confirmation of malignancy using single-cell low-pass WGS [4–6]. Single-cell low-pass WGS characterizes genome-wide copy number alterations (CNAs) that are nearly ubiquitous in solid tumors but occur sporadically in benign tissues [7–9]. Multiple cells with concordant CNA profiles provide compelling evidence of cell malignancy and are thereby identified as genuine CTCs, leading to definitive cancer diagnosis [10].

2 Materials and Reagents

2.1 Reagents

1. HK2 monoclonal antibody, #ab209847, Abcam.
2. Pan Cytokeratin Monoclonal Antibody (AE1/AE3), eFluor 570, #41-9003-82, Thermo Fisher Scientific.
3. Alexa Fluor 488-conjugated goat-anti-rabbit secondary antibody, #A11008, Thermo Fisher Scientific.
4. APC-conjugated CD45 monoclonal antibody, #17-0459-42, Thermo Fisher Scientific.
5. Dithiobis (succinimidylpropionate) (DSP), #22585, Thermo Fisher Scientific.
6. Succinimidyl 3-(2-pyridyldithio)propionate (SPDP), #21857, Thermo Fisher Scientific.
7. Anhydrous DMSO, #D12345, Thermo Fisher Scientific.
8. RosetteSep™ CTC Enrichment Cocktail Containing Anti-CD36, #15167, STEMCELL Technologies.
9. Lymphoprep™ (1.077 g/mL), #18060, STEMCELL Technologies.
10. DAPI (1× solution), #C1006, Beyotime Biotechnology.
11. BSA (Bovine Serum Albumin), #B2064, Sigma.
12. NEBNext Ultra™ II Q5 Master Mix, #M0544, New England Biolabs.
13. Qubit dsDNA HS Assay Kit, #Q32854, Invitrogen.

14. Red blood cell lysing buffer, #555899, BD Biosciences.
15. Agencourt AMPure XP beads, #A63881, Beckman Coulter.
16. Normal Goat Serum, #E510009-0100, Sangon Biotech.
17. UltraPure 1 M Tris-HCl pH 7.5, #15567027, Thermo Fisher Scientific.
18. UltraPure distilled water, #10977023, Thermo Fisher Scientific.
19. PBS, #E607008-0500,Sangon Biotech.
20. Protease, #19155, QIAGEN.
21. Tn5 Transposase, #M045, Novoprotein.
22. 5 M NaCl, #AM9760G, Thermo Fisher Scientific.
23. 0.5 M EDTA (pH 8.0), #AM9260G, Thermo Fisher Scientific.
24. Triton X-100, #A110694-0100,Sangon Biotech.
25. 100 mM DTT, #Y00147, Thermo Fisher Scientific.
26. 10% SDS, #S1010, Shanghai solarbio Bioscience & Technology Co., LTD.
27. 100 μM Cell Strainers, #3523601, Corning.
28. SepMate™-50, #85450, STEMCELL Technologies.
29. Poly-L-Lysine glass slides, #P4981-001, Epredia.
30. K2EDTA Blood Collection Tube, #367863, BD.
31. Fetal Bovine Serum, #A5256701, Thermo Fisher Scientific.
32. Oligonucleotides.

Name	Sequence (5′ → 3′)
Tn5-adaptor1	TCGTCGGCAGCGTCAGATGTGTATAAGAGACAG
Tn5-adaptor2	GTCTCGTGGGCTCGGAGATGTGTATAAGAGACAG
Tn5-ME	CTGTCTCTTATACACATCT
N5 primer	AATGATACGGCGACCACCGAGATC TACACNNNNNNNNTCGTCGGCAGCGTC
N7 primer	CAAGCAGAAGACGGCATACGAGATNNNNNNNNGTC TCGTGGGCTCGG

2.2 Solution Preparation

1. DSP/SPDP cell fixation solution.

Dissolve DSP and SPDP separately in anhydrous DMSO to prepare 50 mM stock solutions, which can be stocked in −80 °C for up to 1 year. Prepare the DSP/SPDP cell fixation solution fresh before use by mixing appropriate volumes of DSP and SPDP stock solutions and diluting with PBS to a final concentration of 5 mM.

2. Tris-HCl (100 mM).

Add 9 mL ultrapure water to 1 mL Tris-HCl (1 M, pH 7.5) to prepare a 100 mM Tris-HCl buffer.

3. Red blood cell (RBC) lysing buffer.

Add 45 mL ultrapure water to 5 mL 10 × RBC lysing buffer to prepare a 1 × RBC lysing buffer (*see* **Note 1**).

4. Blocking solution.

Add 3 g BSA and 10 mL normal goat serum to PBS, mix thoroughly, and adjust the final volume to 100 mL with PBS.

5. Single-cell lysis buffer.

Add 100 μL Tris-HCl (0.1 M), 20 μL EDTA pH 8.0 (0.5 M), 100 μL Triton X-100 (3%), 4 μL NaCl (5 M), 150 μL DTT (0.1 M), 100 μL proteinase (10 mg/mL), and 526 μL H_2O to a 1.5 mL centrifuge tube and vortex gently.

6. Tn5 transposome assembly.

Anneal Tn5-ME with Tn5-adaptor1/adaptor2 oligonucleotides at equimolar ratios using a gradual cooling process (75 °C → 25 °C over 75 min). Mix 4 μL annealed adaptors with 20 μL Tn5 transposase (10 mM), and incubate at room temperature (RT) for 1 h. Store assembled Tn5 transposome at −20 °C.

3 Methods

3.1 Blood and Non-blood Body Fluid Sample Collection and Processing

3.1.1 Peripheral Blood Sample Collection

1. Collect 5 mL of peripheral blood in EDTA vacuum blood collection tube and transport to the lab at 4 °C within 4 h.
2. Add 25 μL CTC enrichment cocktail (5 μL per milliliter blood) to the blood sample, mix gently, and incubate for 10 min at RT.
3. Add 15 mL density gradient medium to a SepMate™-50 cell isolation tube through the hole in the center of the tube insert.
4. Dilute the blood sample with three volumes of PBS-2% Fetal Bovine Serum solution and mix gently.
5. Pipette the diluted blood sample carefully over the 15 mL of density gradient medium in the cell isolation tube.
6. Centrifuge at 1200× *g* for 10 min, brake on (*see* **Note 2**).
7. Discard the top ~10 mL of supernatant. Transfer the remaining ~10 mL of liquid above the barrier to a new 15 mL tube. Add two volumes of PBS-2% Fetal BovineBovine Serum solution and mix gently.

8. Centrifuge at 600× *g* for 8 min, brake low.
9. Discard the supernatant. Add 1 mL of RBC lysing buffer to resuspend cell pellet.
10. Transfer the cells into a new low-bind 1.5 mL tube and lyse cells at RT for 5 min.
11. Centrifuge at 300× *g* for 5 min to pellet nucleated cells.
12. Discard the supernatant. Resuspend cells in 0.5 mL PBS and mix with 0.5 mL DSP/SPDP in PBS to achieve a final DSP/SPDP concentration of 2.5 mM for cell fixation (*see* **Note 3**).
13. Transfer all cells to a Poly-L-Lysine glass slide (*see* **Note 4**). Load cells on the slide surface and fix cells for 40 min (DSP/SPDP) at RT.

3.1.2 Non-Blood Body Fluid Sample Collection and Processing

We use ascites as an example of non-blood body fluid.

1. Collect 10 mL of ascites from the patient and transport to the lab at 4 °C within 4 h.
2. Filter the 10 mL ascites sample through a 100 μm cell strainer.
3. Centrifuge at 300× *g* for 10 min to pellet cells.
4. If red blood cells are present in the sample, treat the sample with RBC lysing buffer to remove them.
5. Wash cells twice with PBS-2% FBS solution.
6. Centrifuge at 300 × *g* for 10 min to pellet cells.
7. Resuspend cells in 0.5 mL PBS and mix with 0.5 mL DSP/SPDP in PBS to achieve a final DSP/SPDP concentration of 2.5 mM for cell fixation.
8. Transfer all cells on the slide with Poly-L-Lysine. Load cells on slide surface and fix cells for 40 min (DSP/SPDP) at RT.

3.2 CTC Enrichment and Detection in Blood and Non-Blood Body Fluids

1. Stop cells fixation by adding 100 μL of 100 mM Tris-HCl (pH 8.0) and incubate at RT for 10 min.
2. Wash the cells on the slides with PBS, three times.
3. Block the cells on the slides with 3% BSA –10% Normal Goat Serum at RT for 1 h.
4. Immunostain the cells on the slides overnight at 4 °C with 200 μL PBS-0.1% BSA containing antibodies (*see* **Note 5**):
   - Anti-HK2 antibody unconjugated (1:100).
   - Anti-pan-CK eFluor 570 (1:200).
   - Anti-CD45-APC (1:50).
5. Wash the cells on the slide with PBS and incubate the cells with Alexa Fluor 488-conjugated secondary antibody (1:400) at RT for 1 h.

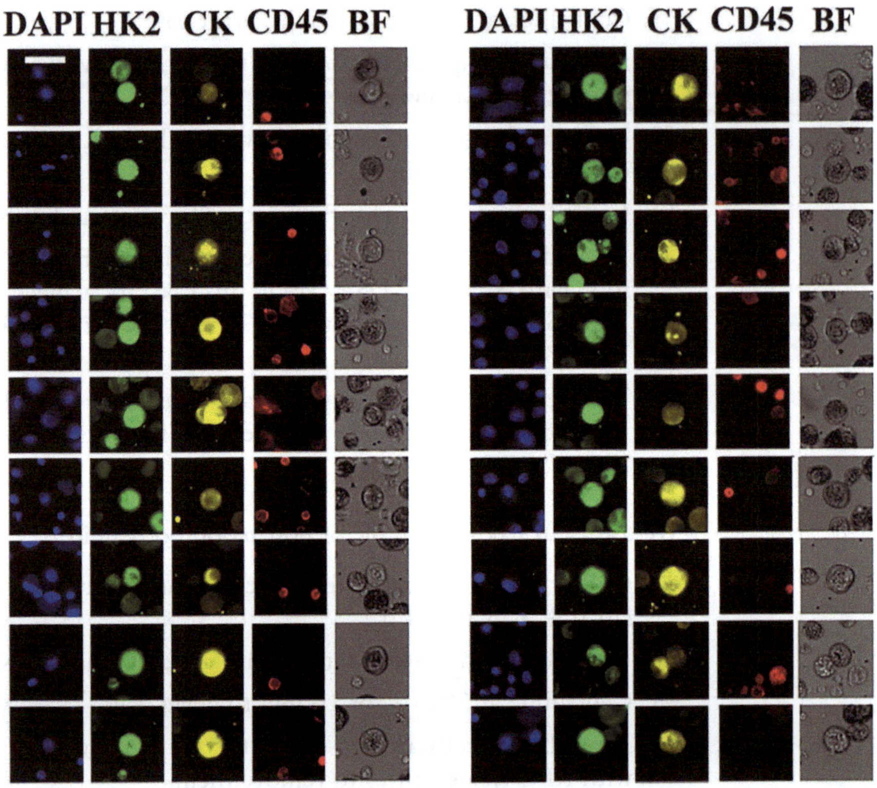

Fig. 1 Representative fluorescence images of CTCs (HK2^{high}/CK^{+}/CD45^{-}/DAPI^{+}) from an ascites sample of a colon cancer patient. Scale bar: 50 μm (reproduced from ref. 5 with permission from Springer Nature)

6. Stain cell nuclei with DAPI at RT for 5 min.
7. Image the cells on the slide using an ImageXpress Micro XLS System (molecular devices).
8. Calculate the threshold of HK2^{high} as the mean HK2 intensity plus 5 times the standard deviation of CD45^{+} leukocytes.
9. Identify suspicious CTCs that exhibit the phenotype of HK2^{high}/CK^{+}/CD45^{-}/DAPI^{+} (Fig. 1).

3.3 Single-Cell WGS

1. Collect suspicious CTCs using a motorized micromanipulator (TransferMan® 4r, Eppendorf) and transfer a single suspicious CTC into a PCR tube containing 2.0 μL of cell lysis buffer.
2. Lyse the cell at 55 °C for 1 h, then inactivate the protease at 85 °C for 15 min.
3. Tagment DNA with 0.125 μL of Tn5 transposome (1:100 dilution) to fragment the genomic DNA and ligate adapters (*see* **Note 6**).
4. Stop tagmentation by adding 0.5 μL 0.1% SDS.

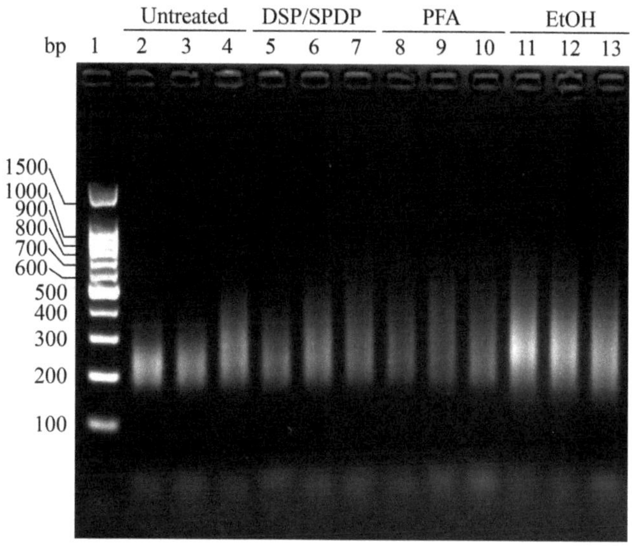

Fig. 2 Agarose gel electrophoresis (2% agarose gel) shows PCR amplified single-cell libraries under different cell fixation methods, including DSP/SPDP, PFA, and ethanol-based cell fixation

5. Perform gap filling at 72 °C for 10 min and then amplify genome sequencing library by NEBNext® Ultra™ II Q5® Master Mix.

   Reaction PCR program:

98 °C	1 min	1 cycle
98 °C	10 s	
60 °C	15 s	20 cycles
72 °C	30 s	

6. Perform agarose gel electrophoresis (2% agarose gel) to check the distribution of PCR amplified single-cell libraries (Fig. 2).
7. Add 18.75 μL (0.75×) of AMPure XP beads to 25 μL of amplified library and mix.
8. Incubate at RT for 8 min, then place sample into magnetic rack for 5 min.
9. Transfer the liquid into a new 200 μL PCR tube, then add 2.5 μL (0.10×) of AMPure XP beads and mix.
10. Incubate at RT for 8 min, then place the sample into a magnetic rack for 5 min.
11. Discard the supernatant, then wash the beads with 80% ethanol twice (*see* **Note 7**).
12. Dry the beads and add 10 μL of ultrapure distilled water to wash purified library.

13. Use a Qubit (Qubit 4 Fluorometer, Thermo Fisher Scientific) to measure the library concentration.
14. Sequence libraries on a NovaSeq 6000 (PE150, *see* **Note 8**).

3.4 CNA Analysis

1. FASTQ files were aligned to the major chromosomes of human (hg19) using BWA (version 0.7.17) with default options.
2. PCR duplicates were removed with Samtools (version 1.11).
3. Aligned reads were counted in fixed bins averaging 500 kb. Bin counts were normalized for GC content with LOWESS regression and bin wise ratios were calculated by computing the ratio of bin counts to the sample mean bin count.
4. The diploid regions were determined using HMMcopy (version 0.1.1).
5. Segmentation was performed with the circular binary segmentation (CBS) method (alpha = 0.0001 and undo.prune = 0.05) from R Bioconductor "DNAcopy" package. Single-cell CNAs profiles were visualized using a R Bioconductor "DNAcopy" package (Fig. 3).

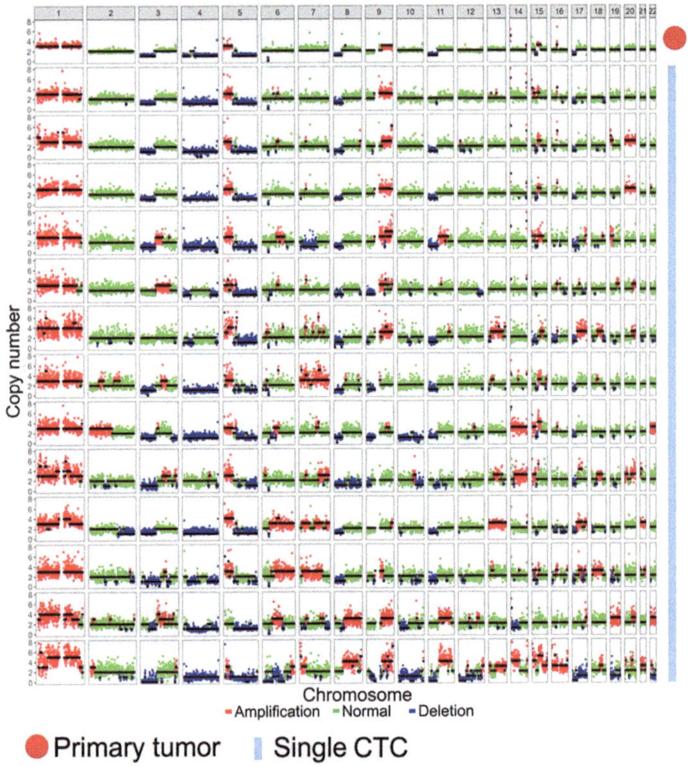

Fig. 3 Single-cell CNAs profiles of 13 single CTCs and the CNAs profile of the bulk tumor tissue from the peripheral blood sample of a small-cell lung cancer patient (reproduced from ref. 5 with permission from Springer Nature)

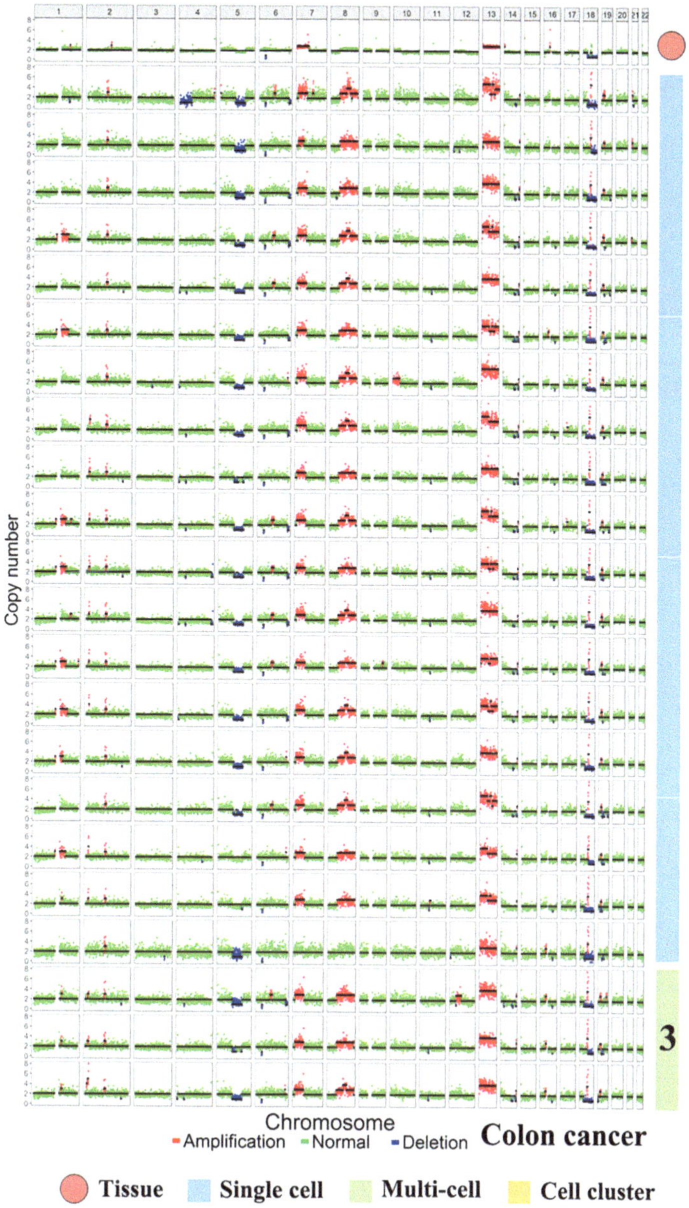

Fig. 4 Single-cell CNAs profiles of 19 single CTCs, 3 three-CTC pools, as well as the bulk tumor tissue from the primary site from the ascites sample of a colon cancer patient (reproduced from ref. 5 with permission from Springer Nature)

6. Copy number noise was quantitated using the mean absolute pairwise difference (MAPD) algorithm. MAPD reliably measures the quality of the amplified genome, and high MAPD scores associate with poor-quality samples. Samples with MAPD ≤0.45 (500 kb bin size) passed the MAPD QC and were included in single-cell genome-wide CNAs analyses.

7. To calculate CNAs burden for a sample, segments of copy number gains and losses are determined, and their total genomic length is summed and calculated as a percentage of the size of the autosomal genome. CNAs burden is defined as the percentage of the tumor autosomal genome with copy number altered.
8. Multiple cells with concordant single-cell CNAs profiles are confirmed as malignant cells and thereby generate cancer diagnosis of the liquid biopsy sample (Fig. 4).

4 Notes

1. Red blood cell lysing buffer should be freshly prepared before use.
2. The centrifugation time can be appropriately extended if the separation efficiency is suboptimal during the density gradient centrifugation of blood samples.
3. In addition to the DSP/SPDP fixation, 4% paraformaldehyde (PFA) fixation and alcohol-based fixation are also applicable for the single-cell WGS protocol.
4. When loading cells onto the glass slides, the cell count should not exceed two million for blood samples and one million for non-blood body fluid samples to prevent excessive cell density.
5. Immunostaining can be performed overnight at 4 °C or for 1 h at 37 °C.
6. Tn5 transposase should be stored at −80 °C and assembled Tn5 transposome can be stored at −20 °C.
7. During magnetic bead purification of single-cell WGS libraries, wash the beads with freshly prepared 80% ethanol.
8. In addition to fluorescence immunostaining, cells from Papanicolaou-stained or Wright-Giemsa-stained cytology slides are also suitable for the single-cell WGS protocol.

Conflict of Interest Statement

The authors have no conflicts of interest to declare that are relevant to the content of this chapter.

References

1. Chaffer CL, Weinberg RA (2011) A perspective on cancer cell metastasis. Science 331: 1559–1564
2. Mohme M, Riethdorf S, Pantel K (2017) Circulating and disseminated tumour cells-mechanisms of immune surveillance and escape. Nat Rev Clin Oncol 14:155–167

3. Mohan S, Chemi F, Brady G (2017) Challenges and unanswered questions for the next decade of circulating tumour cell research in lung cancer. Transl Lung Cancer Res 6:454–472
4. Chen J, Yang Y, Wang Z, Shen X, Zhang Z, Wang C, Xu H, Shi Q (2022) A multicancer malignant pleural effusion diagnostic test using hexokinase 2 and single-cell sequencing. Clin Chem 68:680–690
5. Shen X, Dai J, Guo L, Liu Z, Yang L, Gu D, Xie Y, Wang Z, Li Z, Xu H, Shi Q (2024) Single-cell low-pass whole genome sequencing accurately detects circulating tumor cells for liquid biopsy-based multi-cancer diagnosis. NPJ Precis Oncol 8:30
6. Zhang Z, Liu Z, Chen L, Wang Z, Zhai Y, Qian P, Zhao Y, Zhu L, Jiang H, Wu X, Shi Q (2024) Liquid biopsy-based accurate diagnosis and genomic profiling of hard-to-biopsy tumors via parallel single-cell genomic sequencing of exfoliated tumor cells. Anal Chem 96:14669–14678
7. Zack TI, Schumacher SE, Carter SL, Cherniack AD, Saksena G, Tabak B, Lawrence MS, Zhsng CZ, Wala J, Mermel CH, Sougnez C, Gabriel SB, Hernandez B, Shen H, Laird PW, Getz G, Meyerson M, Beroukhim R (2013) Pan-cancer patterns of somatic copy number alteration. Nat Genet 45:1134–1140
8. Li R, Di L, Li J, Fan W, Liu Y, Guo W, Liu W, Liu L, Li Q, Chen L, Chen Y, Miao C, Liu H, Wang Y, Ma Y, Xu D, Lin D, Huang Y, Wang J, Bai F, Wu C (2021) A body map of somatic mutagenesis in morphologically normal human tissues. Nature 597:398–403
9. Li R, Du Y, Chen Z, Xu D, Lin T, Jin S, Wang G, Liu Z, Lu M, Chen X, Xu T, Bai F (2020) Macroscopic somatic clonal expansion in morphologically normal human urothelium. Science 370:82–89
10. Wang Z, Zhao Y, Shen X, Zhao Y, Zhang Z, Yin H, Zhao X, Liu H, Shi Q (2022) Single-cell genomics-based molecular algorithm for early cancer detection. Anal Chem 94:2607–2614

Chapter 16

Single Oocyst Selection and Whole-Genome Amplification of *Cryptosporidium parvum*

Yuancai Chen, Kaihui Zhang, Jianying Huang, Huikai Qin, Zi Yan, Yin Fu, Junqiang Li, Rongjun Wang, and Longxian Zhang

Abstract

Cryptosporidium parvum is a zoonotic parasite of the intestine and poses a threat to human and animal health. However, it is difficult to obtain a large number of oocysts for genome sequencing using in vitro culture. To address this challenge, we describe a procedure for purification of *C. parvum* from a clinically positive sample and a method for isolation of a single oocyst of *C. parvum* using a three-axis hydraulic micromanipulator. We employ the strategy of whole-genome amplification (WGA) of 10 oocysts, and purify the whole-genome DNA after amplification, and finally meet the basic conditions for long-read sequencing of *C. parvum*.

Key words *Cryptosporidium parvum*, Oocyst purification, Single-oocyst selection, Whole-genome amplification

1 Introduction

Cryptosporidium spp. are parasitic apicomplexans that induce moderate to severe diarrhea in both humans and animals [1]. Currently, nearly 50 species of *Cryptosporidium* have been identified [2]. Several species, such as *C. parvum*, *C. ubiquitum*, and *C. muris*, possess broad host ranges, resulting in zoonotic infections in conjunction with other *Cryptosporidium* species [3]. The absence of broadly effective medications and a lack of a vaccine necessitate a strong emphasis on infection prevention for managing cryptosporidiosis, thereby underscoring the pressing need for innovative interventions [4, 5]. Whole-genome sequencing (WGS) and comparative genomic analysis have been utilized to uncover the genetic factors underlying variations in host range among different *Cryptosporidium* species, as well as the mechanisms of host adaptation within individual species to optimize anti-cryptosporidium drug development and vaccine design [6–8].

Yafeng Ma (ed.), *Whole Genome Amplification: Methods and Protocols*, Methods in Molecular Biology, vol. 3003, https://doi.org/10.1007/978-1-0716-5060-8_16,

One of the major challenges faced in genomics research on *Cryptosporidium* spp. is the scarcity of adequately purified oocysts in sufficient quantities for next-generation sequencing (NGS) analysis, primarily due to the lack of an in vitro culture system capable of propagating the parasites. Previous WGS analyses of *Cryptosporidium* have relied on oocysts purified from laboratory animals that were infected [9–11]. The purity of oocysts obtained by density gradient centrifugation is high, and a large number of oocysts will be lost in the purification process, so it is not suitable for large-scale purification of oocyst samples and is suitable for further purification of oocysts that have been preliminarily gathered [12]. The immunomagnetic bead separation method can detect *Cryptosporidium* specifically, reduce the occurrence of false positive results, and improve the detection accuracy [13]. Flow cytometer technology has the characteristics of a large amount of detection, fast speed, and many measurement indexes [14]. However, the technique of immunomagnetic bead separation and flow cytometer is susceptible to many factors in the actual operation.

Although micromanipulation has not been widely used in the selection of *Cryptosporidium* oocysts, it is one of the best methods to obtain high-purity oocysts. Microoperating systems for manual cell acquisition typically consist of an inverted microscope and a micropipette that can be moved on an electromechanical platform. The micropipette is made of an ultra-thin glass capillary tube that is successively connected to the suction and release units. Cell samples are usually placed in a petri dish or in a specified-sized orifice plate in the form of a suspension. Observed through a microscope, the operator selects specific cells, moves the micropipette close, and sucks the cells by applying suction to the micropipette, and the sucked-out liquid containing the cells can be transferred to the appropriate collection tube. This process is usually performed manually [15]. In recent years, researchers have discovered that there is heterogeneity between cells [16–18], even within the same population; cells can differ at multiple levels (gene transcription and translation, protein activity, metabolite abundance, etc.). With the single-cell isolation method, researchers in related fields can sequence the single cell genome and transcriptome [19].

Troell et al. sequence the genome of a single *Cryptosporidium* oocyst, followed by a thorough whole-genome analysis through comparison with a de novo assembly of the reference population genome. This single-oocyst NGS research marks a significant milestone, demonstrating the feasibility of obtaining high-quality genomic data from single-celled eukaryotes, providing both extensive coverage and precise information [20]. A recent study of *Cryptosporidium* involved only three generations of sequencing of a single oocyst of the genome, not assembly at the chromosomal level.

Before the whole-genome sequencing of *Cryptosporidium*, there were certain requirements for the collection of *Cryptosporidium* oocyst samples, especially the third generation of whole-genome sequencing has higher requirements for oocyst purity. Here, we describe a procedure for purification of *C. parvum* from a clinically positive sample and a method for isolation of a single oocyst of *C. parvum* using a three-axis hydraulic micromanipulator. We employ the strategy of whole-genome amplification of 10 oocysts, and purify the whole-genome DNA after amplification, and finally meet the basic conditions for long-read sequencing of *C. parvum*. It might also be an effective strategy for genome sequencing projects of other difficult-to-collect or uncultivable pathogens.

2 Materials

2.1 C. parvum Positive Stool Sample

1. Strain is isolated from a calf with pre-weaning diarrhea in Henan, China.

2.2 Single Oocyst Purification

1. 80-mesh iron sieve.
2. Funnel.
3. Glass stirring rod.
4. Diethyl ether.
5. Straw tool.
6. Sucrose.
7. Cesium chloride.
8. 15 mL and 50 mL centrifuge tubes.
9. Pipette tool.
10. Centrifugal machine of Thermo-BIRT4i (Thermo Fisher Scientific).

2.3 Single Oocyst Selection

1. Glass petri dish.
2. Microscope (OLYMPUS-BX53).
3. Hydraulic micromanipulator (World Precision Instruments Inc.).
4. Glass capillary tube.
5. Puller PC-100 (Narishige).
6. Microforge MF-900 (Narishige).
7. 4 μL PBS buffer (PBS: 1× PH 7.4; 2.0 mM KH_2PO_4, 137 mM NaCL, 10.0 mM Na_2HPO_4, 2.7 mM KCL).
8. PCR tube.

2.4 Single Oocyst Whole-Genome Amplification

1. REPLI-g® Single Cell Kit (QIAGEN).
2. Lysis Buffer (DTT solution volume: Buffer DLB solution volume = 1: 11).
3. Reaction termination fluid (comes with the kit; the main ingredients are not disclosed).
4. 9 μL H_2O sc (comes with the kit).
5. 29 μL REPLI-g sc (single cell) reaction buffer (comes with the kit; the main ingredients are not disclosed).
6. 2 μL REPLI-g sc (single cell) DNA Polymerase (comes with the kits; the main ingredients are not disclosed).
7. Laboratory water bath.
8. 3 μL Stop Solution (comes with the kits; the main ingredients are not disclosed).

2.5 Whole-Genome DNA Product Purification

1. Agencourt AMPure XP magnetic beads (BECKMAN).
2. Magnetic grate.
3. 80% ethanol (ethanol volume: water volume = 8: 2).
4. 1.5 mL EP tube.

2.6 WGA Product Detection

1. qPCR primer sequence.
   (a) Upstream primer sequence (F): 5′- TAGAGATTG GAGGTTGTTCCT-3′.
   (b) Downstream primer sequence (R): 5′- CTCCACCAAC TAAGAACGGCC-3′.
2. Fluorescent quantitative PCR instrument (qTOWER3 G, Analytikjena).
3. NanoDrop One (Thermo Fisher Scientific).
4. Qubit 3.0 (Invitrogen).

3 Methods

3.1 C. parvum Oocyst Purification

1. Take 4–5 g fresh calf feces sample into a 50 mL centrifuge tube, add water, and stir well with a glass stirring rod. Filter the mixture into a new 50 mL centrifuge tube using a funnel and an 80-mesh iron screen (*see* **Note 1**).
2. Place the above centrifuge tube into the centrifuge (3500 r/min) for 5 min. Use a straw tool to absorb the supernatant and remove it. Add water to the precipitate and stir well with a glass stirring rod. Centrifuge again (3500 r/min) for 5 min. Repeat this **step 3–4** times until the supernatant is clear.
3. Add water to the precipitation in the previous step to 30 mL, and use a glass stirring rod to stir well. Add another 10 mL of

diethyl ether and mix well. Centrifuge (3500 r/min) for 15 min. Remove the diethyl ether layer, lipid layer, and water layer successively with the straw tool (*see* **Note 2**).

4. Add water to the precipitate from the previous step and mix well using a glass stirring rod. Centrifuge (3500 r/min) for 5 min, use a straw tool to absorb the supernatant and remove it. This step is repeated 3–4 times until the odor of diethyl ether in the centrifuge tube is lightened.
5. Prepare 1:1 sucrose solution (1 volume saturated sucrose solution and 1 volume distilled water are evenly mixed), and take 10 mL prepared sucrose solution and put it into a 15 mL centrifuge tube [22]. Dilute the precipitation from the previous step with a small amount of water, and use a glass stirring rod to mix well. Using a straw tool, 2 mL of suspension is drawn and slowly spread on the sucrose solution against the wall of the centrifuge tube.
6. Put the centrifugal tube in the centrifuge (3500 r/min) for 15 min. The oocyst layer is drawn into a new 50 mL centrifuge tube using a straw tool (the liquid after centrifugation appears stratified).
7. Dilute with water and centrifuge (3500 r/min) for 5 min. Use a straw tool to absorb the supernatant and remove it. Repeat this step for 2–3 times.
8. Transfer the liquid containing oocysts concentrate from the previous step to the 1.5 mL centrifuge tube. Take a new 1.5 mL centrifuge tube and add 750 μL cesium chloride solution [23]. Use a pipette tool to absorb 500 μL of liquid containing oocysts, paste the centrifuge tube wall, and slowly spread on the cesium chloride solution.
9. Centrifuge the previous centrifuge tube (13,200 r/min) for 5 min, and use a pipette to draw the oocyst layer into the new 1.5 mL centrifuge tube (the liquid after centrifugation appears stratified).
10. Dilute with water and centrifuge again (13,200 r/min) for 5 min. Use a pipette to remove the supernatant. Repeat this **step 2–3** times.
11. As shown in Fig. 1, concentrate to obtain pure *C. parvum* oocyst fluid [24].

3.2 C. parvum Single Oocyst Selection

1. Use a 10 μL pipette to extract 6 μL of liquid containing pure *C. parvum* oocysts, and place them in a glass petri dish.
2. Place the petri dish on a 60 × inverted microscope stage.
3. A single oocyst of *C. parvum* is isolated using a three-axis hydraulic micromanipulator combined with a glass capillary tube with a diameter of 10 μm (*see* **Note 3**).

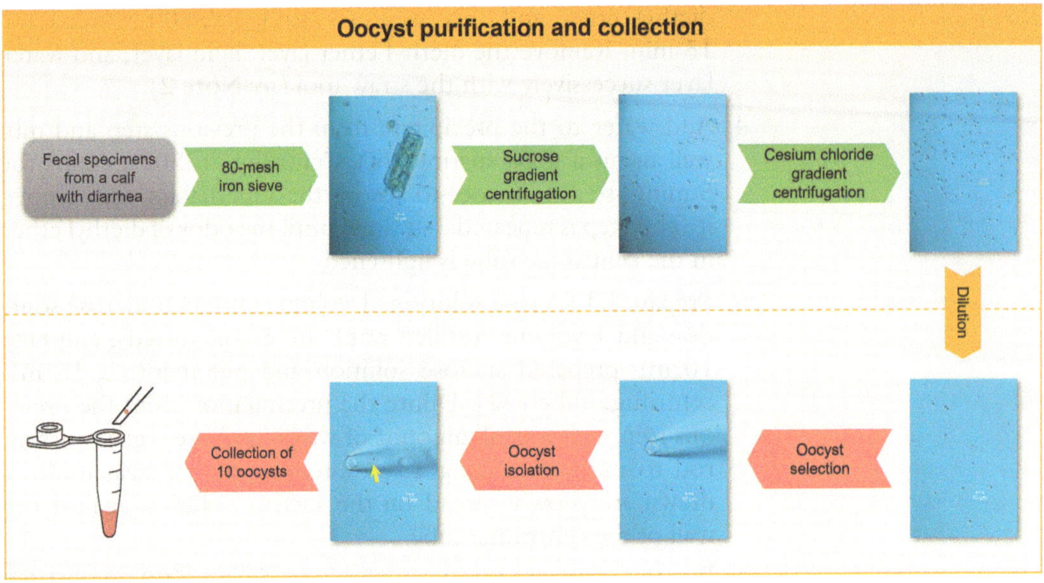

Fig. 1 The purification and collection process of the oocyst (Yellow arrow: *C. parvum* oocyst)

4. After autoclaving, 4 μL PBS buffer is added to the PCR tube.
5. As shown in Fig. 1, 10 oocysts are selected and pooled into a PCR tube containing 4 μL PBS buffer [24].

3.3 Whole-Genome Amplification of a Single Oocyst

Cleavage and whole-genome amplification of *C. parvum* oocysts are performed using the REPLI-g® Single Cell Kit. The basic steps are as follows:

1. 3 μL of Lysis Buffer (Buffer D2; *see* **Note 4**) is added to the PBS containing oocysts, centrifuged for 30 s, and the PCR tube is incubated in a 65 °C water bath for 10 min (*see* **Note 5**).
2. Add 3 μL of the stop solution, centrifuge for 30 s, and place the PCR tube on ice to neutralize the lysate and prevent damage caused by excessive cleavage of the DNA strand.
3. The premixed solution (9 μL H_2O sc, 29 μL REPLI-g sc Reaction Buffer, and 2 μL REPLI-g sc DNA Polymerase; *see* **Note 6**) is prepared. The premixed solution is added into the PCR tube above, and then mix evenly. Centrifuge for 30 s.
4. The PCR tube is incubated in a 30 °C water bath for 8 h (*see* **Note 5**). After DNA amplification, the PCR tube is placed in a 65 °C water bath and incubated for 3 min to inactivate the enzyme.

3.4 Whole-Genome DNA Product Purification

Whole-genome amplification products are purified using Agencourt AMPure XP magnetic beads. Product purification steps are as follows:

1. The magnetic bead solution is oscillated evenly, and 45 μL magnetic bead solution is taken into 50 μL amplified product. Mix well and incubate at room temperature for 5 min.
2. Centrifuge the PCR tube for 30 s and place it on the magnetic grate to separate the magnetic beads and liquid. After 5 min, use a pipette tool to absorb and remove the supernatant, retaining 3 μL of liquid.
3. Prepare fresh 80% ethanol and add 200 μL to the PCR tube. Incubate at room temperature for 30 s, use a pipette tool to absorb and remove the supernatant, and the PCR tube should always be placed in the magnetic rack during rinsing, do not touch the magnetic beads.
4. Repeat **step 3** for two rinses.
5. Open the cover of the PCR tube on the magnetic grate and air-dry the magnetic beads for 10 min until there is no ethanol residue in the PCR tube.
6. Remove the PCR tube from the magnetic grate and add 50 μL double-steaming water for elution. Mix well and let stand at room temperature for 5 min.
7. Centrifuge the PCR tube for 30 s and place it on the magnetic grate for 5 min. Use a pipette tool to draw the supernatant into a new 1.5 mL EP tube.

3.5 WGA Product Detection

1. NanoDrop One is used to detect sample concentrations C (ng/μL), A260/A280, and A260/A230 (*see* **Note 7**).
2. Qubit 3.0 is used to quantify WGA products (*see* **Note 8**).
3. Calculate the Nc/Qc (Nanodrop/Qubit) value (*see* **Note 9**).
4. Take 1 μL of the amplified product and dilute it with double-steaming water at 1: 100.
5. The content of *Cryptosporidium* genomic DNA in diluent is detected by qPCR based on *C. parvum SSU* rRNA gene [25].
6. The reaction conditions are predenaturation at 95 °C for 30 s, followed by 40 cyclic reactions (95 °C for 20 s, 50 °C for 30 s, 72 °C for 30 s).
7. The contents of *C. parvum* genomic DNA in the diluent are characterized by the Ct value obtained by the fluorescent quantitative PCR instrument qTOWER3 G (*see* **Note 10**).

4 Notes

1. To remove large impurities, such as straw stalks. The glass stirring rod, funnel, and 80-mesh iron sieve are disinfection with 100 °C of boiling water before use to keep them clean.

2. The liquid after centrifugation appears stratified. Diethyl ether poses potential harm to both human health and the environment. The waste liquid of the diethyl ether layer is collected and treated in professional containers.
3. Normally, the diameter of the glass capillary tube purchased is 1.5 mm. One end of the glass capillary tube needs to be drawn using the Puller PC-100. The Microforge MF-900 is then used to truncate the glass capillary tube at a diameter of 10 μm.
4. Configure buffer D2 solution (DTT solution volume: Buffer DLB solution volume = 1: 11). The buffer DLB solution needs to be diluted with 500 μL of distilled water before use.
5. Seal the PCR tube with a sealing film to prevent the liquid in the water bath from penetrating into the PCR tube and causing contamination.
6. Create an ice box containing 96 holes, then thaw the DNA polymerase on the ice.
7. The WGA product concentration in *C. parvum* is 700–900 ng/μL. The A260/A280 ratio ranges from 1.8 to 2.0. The A260/A230 ratio ranges from 2.0 to 2.5.
8. Qubit ranges from 25 to 35 μg.
9. The sample quality control is mainly based on Nc/Qc; the closer the ratio is to 1, the higher the purity of the sample and the better the quality.
10. Ct value ranges from 13 to15.

Acknowledgments

This research is funded by the National Key Research and Development Plan Project (2022YFD1800200), NSFC-Henan Joint Fund Key Project (U1904203), and Leading Talents of the Central Plains Thousand Talents Program (19CZ0122). We thank the members of Protist 10,000 Genomes Project (P10K) consortium for their helpful suggestions.

References

1. Kotloff KL, Nataro JP, Blackwelder WC, Nasrin D, Farag TH, Panchalingam S, Wu Y, Sow SO, Sur D, Breiman RF, Faruque AS, Zaidi AK, Saha D, Alonso PL, Tamboura B, Sanogo D, Onwuchekwa U, Manna B, Ramamurthy T, Kanungo S, Ochieng JB, Omore R, Oundo JO, Hossain A, Das SK, Ahmed S, Qureshi S, Quadri F, Adegbola RA, Antonio M, Hossain MJ, Akinsola A, Mandomando I, Nhampossa T, Acácio S, Biswas K, O'Reilly CE, Mintz ED, Berkeley LY, Muhsen K, Sommerfelt H, Robins-Browne RM, Levine MM (2013) Burden and aetiology of diarrhoeal disease in infants and young children in developing countries (the Global Enteric Multicenter Study, GEMS): a prospective, case-control study. Lancet 382(9888):209–222. https://doi.org/10.1016/S0140-6736(13)60844-2

2. Ryan UM, Feng Y, Fayer R, Xiao L (2021) Taxonomy and molecular epidemiology of *Cryptosporidium* and *Giardia* – a 50 year perspective (1971–2021). Int J Parasitol 51(13–14):1099–1119. https://doi.org/10.1016/j.ijpara.2021.08.007
3. Ryan U, Zahedi A, Paparini A (2016) *Cryptosporidium* in humans and animals-a one health approach to prophylaxis. Parasite Immunol 38(9):535–547. https://doi.org/10.1111/pim.12350
4. Bhalchandra S, Cardenas D, Ward HD (2018) Recent breakthroughs and ongoing limitations in *Cryptosporidium* research. F1000Res 7: F1000 Faculty Rev-1380. https://doi.org/10.12688/f1000research.15333.1
5. Chavez MA, White AC Jr (2018) Novel treatment strategies and drugs in development for cryptosporidiosis. Expert Rev Anti-Infect Ther 16(8):655–661. https://doi.org/10.1080/14787210.2018.1500457
6. Fan Y, Feng Y, Xiao L (2019) Comparative genomics: how has it advanced our knowledge of cryptosporidiosis epidemiology? Parasitol Res 118(12):3195–3204. https://doi.org/10.1007/s00436-019-06537-x
7. Khan A, Shaik JS, Grigg ME (2018) Genomics and molecular epidemiology of *Cryptosporidium* species. Acta Trop 184:1–14. https://doi.org/10.1016/j.actatropica.2017.10.023
8. Kim KU, Park SK, Kang SA, Park MK, Cho MK, Jung HJ, Kim KY, Yu HS (2013) Comparison of functional gene annotation of *Toxascaris leonina* and *Toxocara canis* using CLC genomics workbench. Korean J Parasitol 51(5):525–530. https://doi.org/10.3347/kjp.2013.51.5.525
9. Abrahamsen MS, Templeton TJ, Enomoto S, Abrahante JE, Zhu G, Lancto CA, Deng M, Liu C, Widmer G, Tzipori S, Buck GA, Xu P, Bankier AT, Dear PH, Konfortov BA, Spriggs HF, Iyer L, Anantharaman V, Aravind L, Kapur V (2004) Complete genome sequence of the apicomplexan, *Cryptosporidium parvum*. Science 304(5669):441–445. https://doi.org/10.1126/science.1094786
10. Widmer G, Lee Y, Hunt P, Martinelli A, Tolkoff M, Bodi K (2012) Comparative genome analysis of two *Cryptosporidium parvum* isolates with different host range. Infect Genet Evol 12(6):1213–1221. https://doi.org/10.1016/j.meegid.2012.03.027
11. Xu P, Widmer G, Wang Y, Ozaki LS, Alves JM, Serrano MG, Puiu D, Manque P, Akiyoshi D, Mackey AJ, Pearson WR, Dear PH, Bankier AT, Peterson DL, Abrahamsen MS, Kapur V, Tzipori S, Buck GA (2004) The genome of *Cryptosporidium hominis*. Nature 431(7012): 1107–1112. https://doi.org/10.1038/nature02977
12. Entrala E, Molina-Molina J, Rosales-Lombardo M, Sánchez-Moreno M, Mascaró-Lazcano C (2000) *Cryptosporidium parvum*: oocysts purification using potassium bromide discontinuous gradient. Vet Parasitol 92(3): 223–226. https://doi.org/10.1016/s0304-4017(00)00315-0
13. Hadfield SJ, Pachebat JA, Swain MT, Robinson G, Cameron SJ, Alexander J, Hegarty MJ, Elwin K, Chalmers RM (2015) Generation of whole genome sequences of new *Cryptosporidium hominis* and *Cryptosporidium parvum* isolates directly from stool samples. BMC Genomics 16(1):650. https://doi.org/10.1186/s12864-015-1805-9
14. Power ML, Shanker SR, Sangster NC, Veal DA (2003) Evaluation of a combined immunomagnetic separation/flow cytometry technique for epidemiological investigations of *Cryptosporidium* in domestic and Australian native animals. Vet Parasitol 112(1–2):21–31. https://doi.org/10.1016/s0304-4017(02)00414-4
15. Gross A, Schoendube J, Zimmermann S, Steeb M, Zengerle R, Koltay P (2015) Technologies for Single-Cell Isolation. Int J Mol Sci 16(8):16897–16919. https://doi.org/10.3390/ijms160816897
16. Birbrair A (2019) Stem cells heterogeneity – novel concepts preface. Adv Exp Med Biol 1123:V–Vi
17. Dominguez-Gutierrez G, Xin Y, Gromada J (2019) Heterogeneity of human pancreatic β-cells. Mol Metab 27S(Suppl):S7–S14. https://doi.org/10.1016/j.molmet.2019.06.015
18. Kadouri N, Nevo S, Goldfarb Y, Abramson J (2020) Thymic epithelial cell heterogeneity: TEC by TEC. Nat Rev Immunol 20(4): 239–253. https://doi.org/10.1038/s41577-019-0238-0
19. Gierahn TM, Wadsworth MH, Hughes TK, Bryson BD, Butler A, Satija R, Fortune S, Love JC, Shalek AK (2017) Seq-Well: portable, low-cost RNA sequencing of single cells at high throughput. Nat Methods 14(4):395–398. https://doi.org/10.1038/nmeth.4179
20. Troell K, Hallström B, Divne AM, Alsmark C, Arrighi R, Huss M, Beser J, Bertilsson S (2016) *Cryptosporidium* as a testbed for single cell genome characterization of unicellular eukaryotes. BMC Genomics. https://doi.org/10.1186/s12864-016-2815-y
21. Shaw S, Cohn IS, Baptista RP, Xia G, Melillo B, Agyabeng-Dadzie F, Kissinger JC, Striepen B (2024) Genetic crosses within and between

species of *cryptosporidium*. Proc Natl Acad Sci USA 121(1):e2313210120. https://doi.org/10.1073/pnas.2313210120

22. Heyman MB, Shigekuni LK, Ammann AJ (1986) Separation of *Cryptosporidium* oocysts from fecal debris by density gradient centrifugation and glass bead columns. J Clin Microbiol 23(4):789–791. https://doi.org/10.1128/jcm.23.4.789-791.1986
23. Kilani RT, Sekla L (1987) Purification of *Cryptosporidium* oocysts and sporozoites by cesium chloride and Percoll gradients. Am J Trop Med Hyg 36(3):505–508. https://doi.org/10.4269/ajtmh.1987.36.505
24. Chen Y, Huang J, Qin H, Zhang K, Fu Y, Li J, Wang R, Chen K, Xiong J, Miao W, Wang G, Zhang L (2024) Chromosome-level genome assembly of *Cryptosporidium parvum* by long-read sequencing of ten oocysts. Sci Data 11(1):1287. https://doi.org/10.1038/s41597-024-04150-x
25. Cai X, Woods KM, Upton SJ, Zhu G (2005) Application of quantitative real-time reverse transcription-PCR in assessing drug efficacy against the intracellular pathogen *Cryptosporidium parvum* in vitro. Antimicrob Agents Chemother 49(11):4437–4442. https://doi.org/10.1128/AAC.49.11.4437-4442.2005

Index

A

Ampli1 .... 52, 53, 63, 65
Amplification 11, 25, 37, 59, 72, 84, 93, 123, 132, 150, 167, 176, 188, 209, 217

B

Bovine 2, 4, 53, 83, 84, 90–92, 158, 204–206

C

Cell isolation 16, 94, 206
Circulating tumor cells (CTCs) 1, 2, 5, 51, 52, 57, 58, 60, 63, 65–67, 203–211
Clinical metagenomics (CMg) 175, 176
Copy number alterations (CNAs) 204, 210–212
Cryptosporidium parvum 215–222

D

De-branching 178, 180, 185
Φ29 DNA polymerase 132, 134
DNA profiles 37, 38, 45, 127

E

Embryo biopsies 84, 86, 87, 89, 188, 194, 195
Embryos 71, 73, 83–89, 187–191, 194, 195, 197

F

Forensic 37, 38, 44, 48
Formalin-fixed paraffin-embedded (FFPE) 167, 168, 172

G

Gametophyte stages 120
Genome 11, 26, 52, 72, 91, 95, 119, 131, 149, 166, 176, 188, 204, 216
Genomic DNA (gDNA) 13, 16, 20, 34, 38, 42, 46, 73, 75, 79, 90–93, 104, 114, 132, 184, 188, 197–200, 208, 221
Genomic selection 83, 84
Genotyping 11, 12, 38, 83, 84, 87, 90–92

H

Host-depletion 175–182
Hybrid-capture 94, 98, 105, 106, 111, 112, 116–118

I

In vitro fertilization (IVF) 187, 189, 194

L

Laser microdissection (LMD) 26, 27, 33, 165–173
Liquid biopsy 1, 51, 203, 212
Low input tailing method 26
Low template DNA 179, 199

M

Maternal blood 12
Metagenomics 132, 136, 137, 182
Micromanipulation 25, 27, 33, 38, 39, 41–43, 45, 46, 52, 58, 59, 89, 216
Monogenic diseases 187, 188
Multiple annealing and looping-based amplification cycles (MALBAC) 12, 25–28, 31, 32, 35, 72, 79
Multiple displacement amplification (MDA) 25, 38, 48, 84, 93, 103, 113, 121, 127, 128, 131–145, 176, 177
Multiple-strand displacement amplification (MDA) 37–48

O

Oocyst purification 217, 218
Oxford Nanopore Technology (ONT) 149–160

P

Pathogen detection 165, 175
PGT for aneuploidy (PGT-A) 71, 72, 187
PGT for monogenic disorders (PGT-M) 71, 72, 187
Polymerase chain reaction (PCR) 3, 25, 38, 51, 73, 93, 122, 132, 150, 165, 176, 188, 208, 217
Precision medicine 1
Preimplantation genetic testing (PGT) 71, 72

Yafeng Ma (ed.), *Whole Genome Amplification: Methods and Protocols*, Methods in Molecular Biology, vol. 3003, https://doi.org/10.1007/978-1-0716-5060-8,

Primer-independent MDA (piMDA)................. 134, 141
Primer-independent PolB (piPolB).................... 134, 137

Q

Quality control (QC)................... 25–28, 31, 34, 52, 53, 62, 63, 103, 107, 112, 117, 134, 135, 211, 222

R

Rapid barcoding.................................150, 152, 156, 181
Respiratory syncytial virus (RSV)................149–160, 162

S

SeqPlex 167, 168, 170–173
Short tandem repeat (STR).........................38, 44, 46–48
Shrimp viruses 165, 166
Single-cell analysis 1, 94
Single-cell DNA sequencing 51
Single-cell genomics.................................. 11, 12, 51, 93
Single cell sequencing 25, 51, 120, 203–212
Single-cell WGA....... 11, 40, 42, 51, 165–173, 179, 182
Single circulating trophoblast..................................11, 16
Single-oocyst selection 215–222
Single sperm cells 37, 38, 42, 43
Sorghum pollen.. 119–128

W

Whole genome amplification (WGA) 6, 11, 26, 38, 51, 73, 85, 96, 121, 132, 166, 176, 195, 221
Whole genome sequencing (WGS).................. 12, 21, 94, 149–160, 203–211, 215, 217

X

Xenografts.. 51

MIX
Papier aus verantwortungsvollen Quellen
Paper from responsible sources
FSC® C105338

If you have any concerns about our products,
you can contact us on
ProductSafety@springernature.com

In case Publisher is established outside the EU,
the EU authorized representative is:
Springer Nature Customer Service Center GmbH
Europaplatz 3, 69115 Heidelberg, Germany

Printed by Libri Plureos GmbH
in Hamburg, Germany